U.S. MARINES IN IRAQ, 2003 BASRAH, BAGHDAD AND BEYOND

U.S. Marines in the Global War on Terrorism

by
Colonel Nicholas E. Reynolds
U.S. Marine Corps Reserve (Retired)

History Division
United States Marine Corps
Washington, D.C.
2007

Other Publications in the Series
U. S. Marines in the Global War on Terrorism

U.S. Marines in Iraq, 2003: Anthology and Annotated Bibliography

PCN 106 0000 0200

Foreword

The leader of the Chinese communist revolution, Mao Zedong, was once asked by a journalist what he thought was the lasting impact of the French Revolution. He allegedly responded that he did not know the answer to this question as it was "too early to tell." In this same vein, field historian Colonel Nicholas E. Reynolds' book on the beginning of hostilities in Iraq is one of the first historical works commissioned by the History Division to focus on the role of the U.S. Marine Corps in the long war against global terrorism.

This particular book is about Marines during the first stage of Operation Iraqi Freedom (OIF). It spans the period from 11 September 2001 to March and April 2003, when the Coalition removed Saddam Hussein from power, and concludes in November 2003 when the Marines left Kuwait to return to their home bases in the United States. While many then believed that the "kinetic" phase of the fighting in Iraq was largely over, as we now know, it was only a prelude to a longer but just as deadly phase of operations where Marines would be redeployed to Iraq in 2004 to combat insurgents (both foreign and domestic) who had filtered back into the country. However, this phase of the fighting would be very different from the one the Marines and U.S. Army had fought in the spring of 2003 in the march up to take Baghdad.

The primary focus of the book is I Marine Expeditionary Force (I MEF)—the run-up to the war in 2002 and early 2003, especially the development of "the plan," with its many changes, the exhaustive rehearsals, and other preparations, and then the conduct of decisive combat operations and the immediate postwar period, mostly under the control of the U.S. Central Command's Coalition Forces Land Component Command. The book also touches upon other Marine activities in the Military Coordination and Liaison Command in northern Iraq and with the British in the south. Nonetheless, the primary focus remains on I Marine Expeditionary Force and the interactions of its constituent elements. Other forthcoming History Division publications will soon offer detailed narratives on Marines in Operation Enduring Freedom (OEF) in Afghanistan and II MEF operations inside Iraq.

This book is not intended to be the final story on U.S. Marine Corps involvement in Iraq. To paraphrase Chairman Mao, it is too early for that. But it is not too early for a first cut at the role of the Marine Corps in the early phrases of the war in Iraq. These are the first salvoes, intended to bracket the target and start the process of adjusting fire. My view is that it is important to get rounds down range early on, at a time when memories are still relatively fresh and a reasonable number of official sources have become available. My hope is that this book will prove to be a useful overview and introduction to the subject, especially for its Marine students who want to understand the prologue to the continuing war; that it will stimulate further research and healthy debate; and that its readers will perhaps come forward with their own comments and perspectives to be possibly incorporated in follow-on histories.

The author, Colonel Nicholas E. Reynolds, is an infantry officer who has served as a field historian and writer with the Division since 1992. He is the author of two other histories, *Just Cause: Marine Operations in Panama, 1988-1990,* and *A Skillful Show of Strength: U.S. Marines in the Caribbean, 1991-1996*. In January 2000 he became officer-in-charge of the Field History Branch and directed the mobilization and deployment to Iraq of all of the Division's field historians and one combat artist. Colonel Reynolds deployed to the theater to serve with the Military History Group at Coalition Forces Land Component Command. From that base he supervised Marine history operations during the combat phase of Operation Iraqi Freedom and helped to record combined and joint history. Following his deployment, he remained on active duty to write this text and is now a civilian on the faculty at the Naval War College in Newport, Rhode Island. He holds a doctorate in history from Oxford University.

Dr. Charles P. Neimeyer
Director of Marine Corps History

Preface

As pointed out in the foreword, this is not a finished official history, but an operational history from one field historian's point of view. It is not intended to be a desktop guide for future operators, but rather a framework for understanding Marine participation in the Iraq War by describing how Marines coped with the set of challenges they faced in 2002 and 2003. Some of the implied lessons may turn out to be universal, such as the ways Marine staff officers worked in the joint arena. But some of the issues that were pressing in Iraq in March and April 2003 may turn out to be irrelevant in other times and places, except in the sense that past is prologue, that we need to understand where we have been if we want to understand where we are.

This was not a bad time to write this kind of history. I had access to an array of sources that flowed into the then History and Museums Division between the summer of 2003 and the spring of 2004. The efficiencies of the computer age played an important role. It has probably never been easier to track down sources and conduct follow-on interviews or ask participants to expand on the record. Above all, I tried to rely on primary sources—oral history interviews, command chronologies, personal journals, and various contemporary documents. Some of the material is based on personal observation from my time in Kuwait and Iraq in the spring of 2003. I found one or two published accounts particularly useful, especially those that clearly laid out how the author knew what he was reporting. A personal favorite is *In the Company of Soldiers* by Rick Atkinson, a journalist who is also a respected military historian. I have written extensive endnotes in the hope that future students of the war will be able to continue where I leave off. Occasionally, I have purposely omitted a name or a source because of the sensitivity of what someone told me. But that is rare. While I am not as free with my opinions as a journalist or an academic might be—I was, after all, writing on government time as a member of a disciplined Service—I have tried to report both the good and the not so good.

This monograph would not have been possible without the assistance of a lot of people. I want to begin by acknowledging my brother Marines, and one sister Marine, in the Field History Branch, who, in 2003, did an amazing job of collecting the raw material of history before it was lost. They served in various billets, both Marine and joint, from the Horn of Africa to northern Iraq, enduring austere and sometimes dangerous conditions to get the job done. Their examples often inspired me. Lieutenant Colonel David T. Watters, my successor as officer-in-charge, was the first to volunteer for mobilization in the fall of 2002, when the war clouds were still forming; Major Melissa Kuo, who served with 1st Force Service Support Group; Chief Warrant Officer William E. Hutson, who drew a difficult assignment in the early days of the war, wrote to me to thank me for giving him the opportunity to serve instead of complaining about his luck; Colonel Reed R. Bonadonna, a talented journal writer, was with Task Force Tarawa during the heavy fighting at An Nasiriyah; Lieutenant Colonel Michael D. Visconage and Major Carroll N. Harris were at neighboring commands during the war, and we formed a kind of historical wolf pack when we flew into Iraq in late April to conduct a series of interviews in Ad Diwaniyah. It was a week I will not soon forget. Major Theodore R. McKeldin III, Lieutenant Colonel Tommy Ryan, USA, and I visited with the British division a few days later, collecting data in and around Basrah. Our British colleagues were wonderful hosts, and I cannot say enough good things about their hospitality and their openness. My hosts at Coalition Forces Land Component Command in the Military History Group under Colonel Neil Rogers, USA, were equally generous with their resources, especially

Major Shane Story, USA, who stayed in touch after the war while he was writing the command's official history.

After returning to Washington, D.C., I read every one of our historians' personal journals, as well as many of the documents and oral histories they collected, great sources for putting some flesh on the dry bones of official records. For the same reason, I looked at a lot of great combat art created by Staff Sergeant Michael D. Fay who is not only one hell of a field Marine, but also the only other member of the branch who was over 50 when the war broke out. It was good to have at least one other Marine in the group who could remember what it was like to be in the Corps in 1975. All of the field historians' materials ultimately found their way into the unit's finding aid, ably compiled by Lieutenant Colonel Nathan S. Lowrey to provide a guide to the vast amounts of data that we collected. Colonel Jeffrey Acosta, who served at Marine Corps Forces Central Command headquarters in Bahrain during the war, added many useful documents he collected, as well as a short history of the command in the operation. Colonel Jon T. Hoffman, Major Christopher J. Warnke, and Captain Christopher M. Kennedy did the same after their trips to Iraq in the late spring and summer of 2003.

I also want to express my appreciation to all of the Marines who received our field historians in March and April 2003. The 3d Marine Aircraft Wing and the 1st Force Service Support Group went out of their way to make us feel welcome. So too did virtually all of the Marines who sat for interviews, before and after the war. They interrupted busy schedules to meet, in some cases for many hours. One commander even sat for a video teleconference interview a few hours before his Marine expeditionary unit mounted out for the Middle East. Another commander met with me for half a day in Quantico, Virginia, during his pack out to return to I Marine Expeditionary Force and ultimately to Iraq. Talk about dedication to the historical record!

The staff of the History and Museums Division was uniformly helpful, from the director, Colonel John W. Ripley, and the deputy director, Colonel Hoffman, to the Marine on the quarterdeck. I want to express my thanks to all and to single out a few people for special thanks. The chief historian of the Marine Corps, Charles D. Melson, was there for us whenever we needed him. During the writing stage, he shared the remarkable trove of data that he collected, and he was good enough to read the manuscript in draft. The Field History Branch partnered with the Oral History and Archives Sections to conduct and access some 1,300 interviews. Dr. Gary D. Solis, friend and colleague, was instrumental in the process, as were Dr. Fred Allison and Frederick J. Graboske and his staff, Christine Laba and Robert Piehel. Danny J. Crawford, head of the Reference Branch, remains one of the linchpins of the Division. Annette Amerman, also of Reference, did great work correcting and editing the troop list. I would be remiss if I omitted Charles Grow, the curator of the art collection, always imaginative, helpful, and cheerful; graphic designer William S. Hill, who produced wonderful maps; librarian Evelyn A. Englander, always willing to chase down one more source for a writer; Charles R. Smith, Major Valerie A. Jackson, Greg Macheak, and Wanda J. Renfrow, who edited and illustrated the book; and Mrs. Carol Beaupre, the director's executive assistant, ever responsive to calls for help of many kinds.

Mr. Jay Hines, the Central Command historian, was good enough to read and comment on the text, as was my neighbor at the Navy War College, Colonel William J. Hartig, late of the I Marine Expeditionary Force staff. The manuscript finally got into decent shape thanks to the world-class word-processing skills of Ms. Tina Offerjost of Stafford, Virginia, not to mention the editing skills of Ms. Jill Hughes.

Last but far from least is my wife, Becky, fondly known as "the Boss" to the branch, who not only agreed to let me interrupt my civilian career one more time and go to war for a few months but was also as loving and supportive as any man could want while I was away. I promised her that this would be my last big adventure. I will try to keep that promise. But I will always cherish the camaraderie of the field historians, an unusual band of brothers, and keep a uniform on hand, just in case.

Nicholas E. Reynolds
Colonel, U.S. Marine Corps Reserve (Retired)

Table of Contents

CAT
F.D.N.Y.

Chapter 1

Prelude to War

Before 11 September 2001, the Marines at U.S. Central Command (CentCom) were housed in a tan building that looked something like a double-wide trailer on cinder blocks. It stood almost literally in the shadow of the imposing, and very permanent-looking, CentCom headquarters at the north end of MacDill Air Force Base near the shores of Tampa Bay in Florida. Up until then it was forgivable for visitors to think that the Marine Forces, Central Command, known as MarCent Tampa, was an afterthought, one of CentCom's many appendixes. That was not quite the case, although the situation was as complicated as an infernal machine designed by Rube Goldberg.

MarCent Tampa was a liaison element run by a chief of staff, Colonel John A. Tempone, who was stationed at MacDill and charged with representing the Marine Corps before CentCom. Colonel Tempone did not command any fighting forces, and there was no commanding general on the books, although there was an informal arrangement whereby the senior Marine on the CentCom staff could step into that role. When there was a formal requirement for Marine forces, or for a Marine general officer, the Commanding General, Marine Forces Pacific (MarForPac), who had two corps-sized Marine expeditionary forces at his disposal, would engage; curiously, MarCent Tampa was an outpost of that command, whose headquarters was thousands of miles away in Hawaii at Camp H.M. Smith on the island of Oahu. One of the few things MacDill and Smith had in common was palm trees. With forces spread from California to Korea and Okinawa, MarForPac was, by name and location, part of Pacific Command, well placed for engagement in Indonesia, Korea, or the Philippines, while CentCom was responsible for the Middle East, on the other side of the world. It was the joint command that had fought the Gulf War of 1990-1991 against Iraq and maintained steady pressure on Iraq throughout the 1990s and into 2001, mostly by orchestrating an international Coalition of forces to enforce the no-fly zones in the northern and southern thirds of the country, which were known as Operations Northern Watch and Southern Watch.[1]

CentCom was now very much in the limelight again. The day after the attacks on New York and Washington on 11 September, Secretary of Defense Donald H. Rumsfeld directed CentCom to prepare "credible military options" to neutralize the terrorist threat to the United States. This spanned contingency plans against a variety of potential targets, including Iraq and Afghanistan. President George W. Bush's administration initially believed these two countries had played a role in supporting or sponsoring the attacks. However, it soon became clear that it was Afghanistan, not Iraq that harbored the terrorist organization that had planned the attacks. By 21 September the CentCom commander, General Tommy R. Franks, USA, had briefed the President on a plan to take the war to that organization, the Moslem fundamentalist Al Qaeda, and to the equally fundamentalist Taliban government of Afghanistan. It would not be an exaggeration to say that Al Qaeda, run by the Saudi millionaire Osama bin Laden, had purchased a share of the government, and that he could more or less do as he pleased in the remote, mountainous country wedged in between Pakistan and the other "stans," the small successor states to the Soviet Union north of Afghanistan.[2]

General Franks' plan to destroy the Taliban and to eradicate Al Qaeda relied heavily on indigenous opposition forces within Afghanistan, especially those known as the Northern Alliance, with heavy support from U.S. airpower, special operations forces, and other government agencies. A broad range of Coalition partners, especially North Atlantic Treaty Organization (NATO) allies, offered various kinds of support. The operation ultimately known as Enduring Freedom would consist of several simultaneous lines of operation ranging from reconnaissance and information operations to unconventional warfare and deep air attacks on enemy lines of communication. American forces would not be heavily committed on the ground. It was, like Goldilocks' porridge, neither too much nor too little, but "just right," an imaginative mix of traditional and "transformational" approaches. No one on the United States side wanted to follow the Russian example, set during years of bloody and ultimately futile fighting, mostly by conventional forces against guerrilla bands that ended in Russian defeat in 1989. Nor did they want to follow earlier American examples of simply firing a few missiles at an elusive enemy.[3]

DVIC DN-SD-03-11500

On 11 September 2001, five members of Al Qaeda hijacked American Airlines Flight 77 shortly after it took off from Dulles International Airport outside Washington, D.C. After following a circuitous route which took them away and then back toward Washington, they flew the aircraft into the side of the Pentagon. The impact destroyed four of the five "rings" in a section of the building, killing 64 on board the plane and 125 on the ground.

Exactly what the Marine Corps could or would contribute to a fight in a land-locked country was not clear at first, but the MarForPac staff in Hawaii began to focus far more on CentCom than it normally did, eventually splitting itself more or less in two, with one group for the Pacific and another for the Middle East. The first challenge was to gain and maintain "situational awareness," jargon in this war for knowing what was going on, and then to develop and weigh potential courses of action. For example, could any of the battalion-sized special operations capable Marine expeditionary units, which were loaded on ships, spread throughout the world, and combat ready, be brought to bear? Two suggestions came from then-Lieutenant General Michael W. Hagee, who was serving as the commander of I Marine Expeditionary Force (I MEF). He passed the word up the chain-of-command to CentCom that the Marine Corps could contribute special operations capable expeditionary units to a fight in Afghanistan and that it could stand up a task force for a mission known as "consequence management."[4]*

This was certainly in line with the thinking of the Commandant of the Marine Corps, General James L. Jones, Jr., who wanted it known that the Marine Corps was willing and able to commit a Marine expeditionary brigade to the new contingency. General Jones was a proponent of the view that a brigade (which could be made up of two expeditionary units) was the kind of organization that worked well in joint and Coalition operations; a brigade of any kind was a relatively familiar concept to a planner from another Service.[5]*

The first suggestion set in motion a chain of events that led to the creation, under MarCent's operational control, of the Combined Joint Task Force Consequence Management (C/JTF-CM), which came to be based at Camp Doha, Kuwait. Commanded and staffed largely by Marines from I Marine Expeditionary Force augmented by Reserve Marines from I MEF Augmentation Command Element (I MACE), C/JTF-CM's unusual mission was to assist local governments in coping with the effects of a nuclear, bi-

---

*A draft of "The Informal History of MarCent" states: "as operations developed, ComUSMarCent was tasked by the Secretary of Defense, and the chairman [of the] Joint Chiefs of Staff and Commander, Central Command to–(1) establish . . . C/JTF-CM [and] (2) support Coalition . . . [forces] with two Marine MEUs under the command of a Marine General Officer . . . [while] deploying MarCent headquarters to Bahrain" (USMarCent, "The Informal History of MarCent," copy in Reynolds Working Papers, MCHC, Quantico, VA)

*Gen James L. Jones contrasted the utility of a Marine expeditionary brigade with that of a Marine expeditionary force-forward, which he saw as less useful in the joint/combined arena. The brigade has a checkered history. It was coming back into favor in the early part of the 21st century. Gen Anthony C. Zinni also noted the continuing struggle between the "MEFers," who believe that the Marine Corps should focus its energies at the force level, and the "MEBers," who believe the same of the brigade. (Gen Anthony C. Zinni intvw, 7Jan04 [MCHC, Quantico, VA], hereafter Zinni intvw) ological, or chemical (NBC) attack by, for example, Al Qaeda or Iraq. Its table of organization was equally unusual, eventually coming to include contingents from a number of allied nations, especially NBC units from Germany, Slovakia, and the Czech Republic. Although it was never called to manage the consequences of an NBC attack during the 18 months of its formal existence, from October 2001 to May 2003, it could have made an enormous difference had Kuwait or another Arab ally been attacked and was a notable experiment.[6]

Thousands of miles away from Kuwait, Afghanistan, or Hawaii, the small MarCent office building in Tampa began to hum with operational intensity–the increasingly more organized and purposeful chaos that characterizes a successful wartime staff. The deputy commander of MarForPac, Brigadier General John G. Castellaw, came to spend time on scene, helping with the interface between that command and CentCom. Other augmentees of various sorts, including recently retired officers and the categories of reservists known as individual mobilization augmentees and individual ready reservists, flowed in.*

Only half in jest, some began to refer to MacDill as "Tampastan." The workday there was now as long as the workday for Americans in or near Afghanistan, but it was more complicated in some ways, and there were still the peacetime responsibilities of home and family after hours. It was not dangerous like combat, but it was very stressful, and a few CentCom officers suffered heart attacks or other forms of burnout. For Marines there was the additional twist of a complicated "battle rhythm" that spanned numerous time zones. They often had to repeat the same evolution many times over. For example, if an event happened in the Middle East during the day, which was nighttime in Tampa, MarCent officers faced it when they came to work in the morning, which was still nighttime in Hawaii. They would then have to recap the facts, and their discussions, for MarForPac a few hours later, near the end of their workday but at the beginning of MarForPac's. On many issues, they also had to engage I MEF, which was a constituent part of MarForPac. Built around 1st Marine Division, 3d Marine Aircraft Wing, and 1st Force Service Support Group, it was the Marine air-ground task force that was most likely to send Marines to fight in the Middle East. Being located at Camp Pendleton near San Diego, California, it was in yet another time zone.

Photo courtesy of U.S. Central Command

An artilleryman by training, Gen Tommy R. Franks, USA, commanded the U.S. Third Army before being selected for promotion to general and assignment as Commander in Chief, United States Central Command. Franks succeeded Marine Gen Anthony C. Zinni to this position on 6 July 2000.

There was another small staff at work on contingency plans for Marine operations against the enemy in Afghanistan, the MarCent coordination element, commanded by the redoubtable Colonel John B. Kiser and based at the Naval Support Activity, Bahrain, a compact facility where the Commander, U.S. Naval Forces, Central Command (NavCent) flew his flag. To bolster the handful of officers at the coordination element, the Marine Corps created the designation "Commanding General, MarCent-Forward" for an officer who was already in theater for the multinational CentCom exercise in Egypt known as "Bright Star," and who was able to assume some of the responsibilities of a Service component commander.* That officer was Brigadier General James

*Some picked up the rhythm more quickly than others. This was true of the retired officers, some with CentCom experience, and the individual mobilization augmentees (IMAs), who were trained for specific wartime jobs, but tended to be less true of the individual ready reservists, who came from a large pool of less active reservists. Although MarForPac had an IMA detachment, quickly mobilized to augment its staff to take on the increased responsibilities of two fronts, the small liaison element at MarCent did not. The Marine Corps' Enduring Freedom Combat Assessment Team suggested that the lesson learned here was, if you do not have your own IMAs you have shaped to your needs, the Marine Corps' personnel system is unlikely to give you what you want in times of crisis, and you should not be surprised when that happens. (U.S. Marine Corps, *Operation Enduring Freedom Combat Assessment Team Summary Report* [Quantico, VA: MCCDC, 2003], p. 73, hereafter MCCDC, *OEF Summary Report*)

*"BGen [James N.] Mattis, originally designated ComMarCent (Forward)... assumed significant responsibilities as the Marine Service component commander forward in the A[rea of] O[perations] for ensuring the proper employment, administration, and sustainment of Marine Corps Operating Forces in theater." (MCCDC, *OEF Summary Report,* p. 63)

N. Mattis, the commanding general of the 1st Marine Expeditionary Brigade, the smaller Marine air-ground task force embedded in I MEF, who was tasked briefly with commanding C/JTF-CM but soon shifted his focus to Bahrain and Afghanistan.[7]

A historian could follow the course of who was in Operations Enduring Freedom and Iraqi Freedom by answering the question, where is General Mattis? Born in Richmond, Washington, and educated at Central Washington State College, Mattis was commissioned on 1 January 1972. In 2001 he was a trim 51-years-old. Especially when wearing his reading glasses, he looked like a military intellectual, which would not have been far from the mark. But he was also a field Marine par excellence. Even in a Corps of energetic men and women with a bias for action, to say nothing of their single-minded devotion to their profession, he stood almost in a class by himself. General Mattis' official biography was characteristically brief; it did not contain any personal data, only a list of the military schools he had attended and the commands he had held, along with the obligatory official photograph, without the glasses. He had been a lifelong student of war, known for his voracious but discriminating appetite for the printed word. He believed "we face NOTHING [emphasis in original] new under the sun." It followed that to understand a problem; the approach he recommended was to study its history. For example, during Desert Shield/Desert Storm, when he had commanded an infantry battalion, he had read Bruce Catton's *Grant Takes Command*, studying the need for commanders to get along, as well as books by Erwin Rommel and Bernard Law Montgomery, presumably to learn about desert warfare.[8] On one occasion, General Mattis described himself as a student of "Sun Zinni," a tongue-in-cheek reference to another "one of them field Marines that reads," the iconoclastic General Anthony C. Zinni, who had retired in 2000 after serving as the commander-in-chief at CentCom and had always stressed the need to understand the cultural dimension of war. General Mattis could relate well to enlisted Marines and to senior Pentagon officials, equally comfortable in the roles of salty platoon commander and, less salty, policy maker. Some have described General Mattis as more demanding than most commanders and slow to warm to officers who were not on his team, but also as willing to "go to hell and back" for his Marines.[9] He liked to say that the secret of success was "brilliance in the basics." There can be no question that that is what he demanded of his Marines, but there was nothing basic about his ability to think outside the box.

A native of Washington state, MajGen James N. Mattis commanded the 1st Marine Expeditionary Brigade and Task Force 58 during Operation Enduring Freedom in Afghanistan before assuming command of the 1st Marine Division. As commander of Task Force 58, he became the first Marine to command a naval task force in combat.

DVIC DM-SD-07-06012

Within days of his arrival in Bahrain on 27 October, General Mattis was shaping plans for amphibious raids into Afghanistan, in effect following up on General Hagee and General Jones' readiness to use a Marine expeditionary unit or, better yet, a Marine expeditionary brigade in that country. Mattis reported to Vice Admiral Charles W. Moore, USN, the commander of NavCent, which included the Marines then afloat in theater. Admiral Moore was charged with responsibilities ranging from deployment and sustainment to warfighting, which meant he was able to send forces into combat. Moore has been described as an unusually aggressive officer, interested in finding ways to take the fight to the enemy; he and General Mattis had no trouble understanding each other. A fellow innovator, Admiral Moore took the unusual step of designating Mattis as the sole commander for the ad hoc Task Force 58, which made him "the first Marine to command a naval task force in wartime."[10] General Mattis was now serving both as ComMarCent-Forward and as commander of a task force

DVIC DM-SD-06-03033

A Marine from the 15th Marine Expeditionary Unit (Special Operations Capable) carries a full combat load, including an FNMI 7.62mm M240 machine gun, while moving into a security position after seizing Forward Operating Base Rhino in support of Operation Enduring Freedom in Afghanistan.

preparing for combat. In accordance with his request, it would be designated Task Force "Chaos" after the effect he wanted to have on the enemy, and he himself would use the call sign "Chaos" for the next two years.[1]

Under the control of NavCent were the two amphibious ready groups (ARGs) in theater. Each carried a Marine expeditionary unit, special operations capable, built around a battalion landing team of infantry, a helicopter squadron, and a combat service support element. The two units were the 15th Marine Expeditionary Unit (Special Operations Capable) (15th MEU(SOC)), in the ARG built around the USS *Pelelieu* (LHA 5), and the 26th MEU(SOC), in the ARG built around the USS *Bataan* (LHD 5). Together they formed Task Force 58. Admiral Moore made it clear he did not want Task Force 58 to have a large staff, like some Marine expeditionary brigades that had literally hundreds of Marines and sailors on their books. This was fine with Mattis, who liked to work with a very small staff of trusted individuals. The upshot was that the staff stayed small, never exceeding 40 officers and men. It focused on broad-brush planning, while the Marine expeditionary units relied on their own staffs for the detailed planning and conduct of operations. A sensible corollary was the decision not to "composite" the two expeditionary units, that is, to meld them into one force under one staff, but to keep them intact and to create a supported/supporting relationship between them whereby one expeditionary unit would take the lead for a time, and then the other. In the words of the Task Force 58's command chronology: "While one MEU executed a mission, the second MEU . . . [c]ould conduct detailed planning for the follow-on mission."[12*]

By 3 November, General Mattis had briefed Admiral Moore on his concept of operations. The admiral provided additional guidance, telling the Marines he wanted them to defeat Taliban and Al Qaeda forces. Moore thought the Marines would make a difference, that a "squad of Marines running through Kandahar would turn the tide."[13] But soon the mission changed from conducting raids to seiz-

*The combat assessment team surmised that this "staff lite" arrangement worked because it included the two traditional Marine expeditionary unit staffs; "staff lite" by itself might not have been as successful. (MCCDC, *OEF Summary Report*, p. 64)

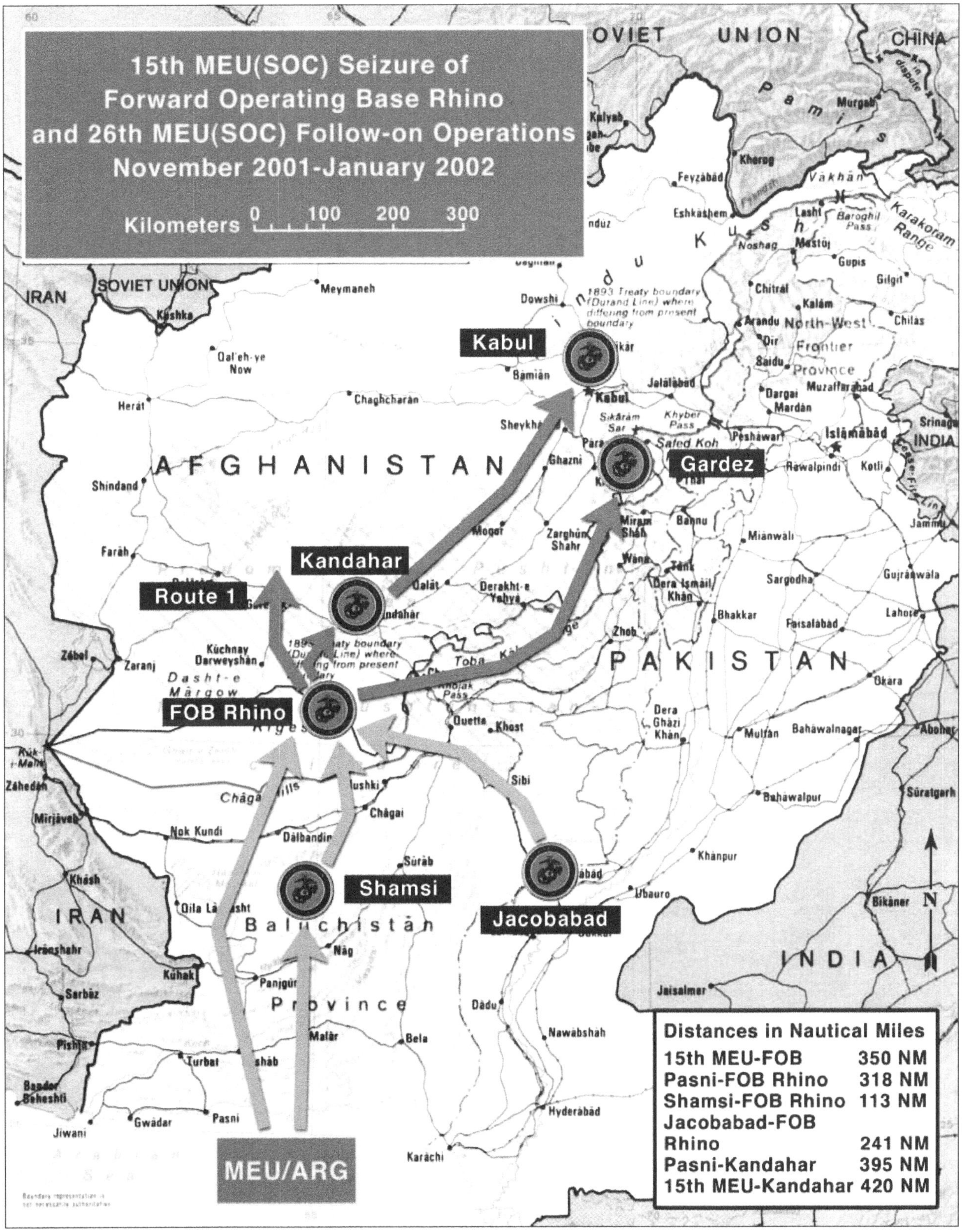

ing a forward operating base in order to attack lines of communication and generally support Coalition operations.[14]

By the middle of November, the Coalition's Afghan allies had driven the Taliban's ragtag forces from most of the country's larger cities, including the capital, Kabul, and Task Force 58 was ready to launch. On 25 November, traveling some 400 nautical miles

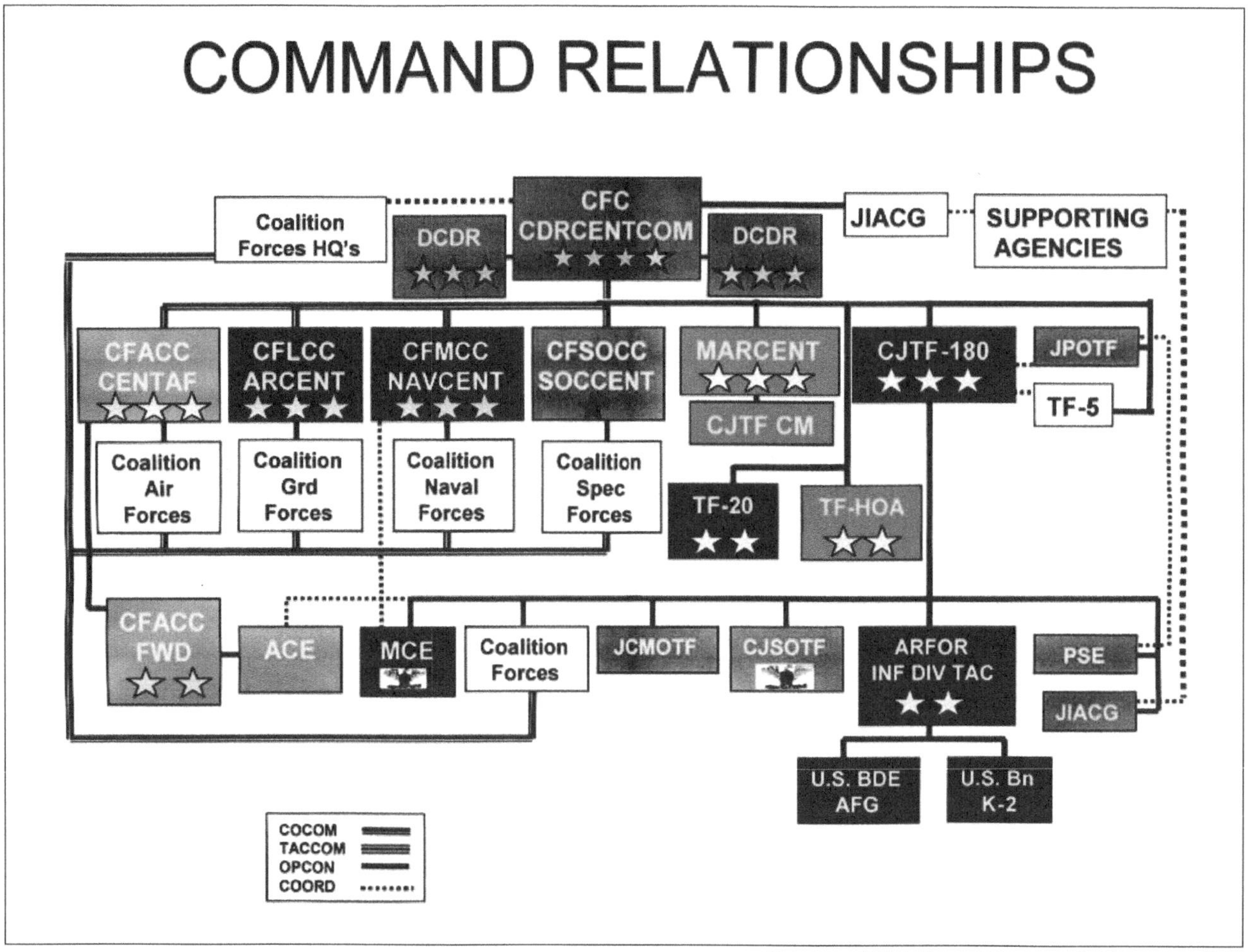

Courtesy of U.S. Central Command, 14 January 2003

from the beachhead, 15th MEU (SOC) flew in to seize a landing strip in Afghanistan that became known as Forward Operating Base Rhino. This was the start of roughly two months of Marine operations in country, first in and around Rhino, and later from Kandahar Airfield, which 26th MEU (SOC) later seized on 14 December.

Repeating a phrase attributed to war correspondent Richard Harding Davis in 1885 and used many times since, General Mattis wrote in a 26 November message, "the Marines have landed and the situation is well in hand."[15] * It was, in the words of the Naval Institute's annual review of the Marine Corps, a display of "flexibility and operational reach . . . [that] stunned many outsiders."[16] The task force's movement from ship to shore was right out of the Marine Corps doctrine known as expeditionary maneuver warfare, which called for fast and deep movements, directly from the sea to objectives inland. But this was still something relatively new and largely untried. General Mattis was not exaggerating when he commented that if he had proposed this kind of operation at Quantico or Newport, Rhode Island, the home of the Naval War College, he would have been told it could not be done.[17]

In the meantime, back in Hawaii there were changes afoot that would shift some of the burden from the small liaison elements in Tampa and Bahrain, and from General Mattis himself. On 24 October, the Commander, MarForPac, Lieutenant General Earl B. Hailston, had formally taken on the additional duty of Commander, MarCent, and begun to play a much more prominent role in the CentCom arena.[18] By January 2002, Hailston was in the process of moving his flag to Bahrain "to establish [the] MarCent HQ in theater in support of CincCent [Commander-in-Chief, CentCom] and Operation Enduring Freedom."[19] Even though he occasionally shuttled back to Hawaii, Bahrain became his home for much of the next year and a half.[20] Hailston's move, made with some 200 of his Marines, signaled the Marine

*Some journalists reported the remark as, "The Marines have landed and we now own a piece of Afghanistan," which the Pentagon apparently "scrambled to disavow." (See, for example, Christopher Cooper, "How a Marine Lost His Command in the Race to Baghdad," *Wall Street Journal*, 5Apr04, p. 1)

Corps' intent to build a strong infrastructure in theater to support Marine commanders on the ground. It was as much a matter of Iraq as it was of Afghanistan; by early 2002 the administration had directed the military to plan for a possible war with Iraq while consolidating its gains in Afghanistan.

The new MarCent headquarters in the Gulf was not even up to the standard of MarCent Tampa's building; like virtually all Marine headquarters in and around the Persian Gulf through the summer of 2003, it was never much to look at, let alone to work in. At first MarCent even had trouble finding a home in the theater. When they landed at Sheikh Issa Air Base in Bahrain, the Marines from Hawaii and Tampa started to build an expeditionary headquarters on a part of the base. But the initiative had not been fully cleared with the Bahrainis, there had been some kind of cross-cultural misunderstanding, and they "requested" that MarCent find somewhere else to go. The Marine coordination element commander, Colonel Kiser, who was familiar with Arab ways and with the Navy establishment, came up with a quick

Componency

The first time outsiders tried to understand the relationship between Central Command's "service components" and "functional components," their heads spun and they had to reach for their favorite painkiller. Each Armed Service had a Service component, with a headquarters and a staff, commanded by its senior officer in theater. These components were MarCent, NavCent, ArCent, and CentAF. Their purpose was to provide and sustain their forces. NavCent and CentAF were commands that could easily transform themselves into functional commands, that is, commands responsible for combat at sea and in the air. It seemed natural for the naval commander to take on the additional duty of the Coalition Forces Maritime Component Commander for naval warfare, while the CentAF commander could become the Coalition Forces Air Component Commander for the fight in the air. There was even a commander for special operations, Coalition Forces Special Operations Command (CFSOC). It also seemed natural for the ArCent commander to become dual-hatted as the Coalition Forces Land Component Commander, but what about MarCent? There were no provisions for an "Expeditionary Forces Component Command." In the Gulf War of 1990-1991, there had been nothing even remotely like a functional component command for the war on land; the CentCom commander, Army General H. Norman Schwarzkopf, had decided to take on the additional duties of being his own land component commander, not unlike the Confederate general Braxton Bragg, who had been his own supply officer. After he became the CentCom commander in the late 1990s, General Zinni had wrestled with the problem. His view was that the United States had fought two ground wars in Desert Storm. General Schwarzkopf had taken on more than any one man could handle, and the efforts of the Army and the Marine Corps had not been fully coordinated. Zinni decided that he wanted the Army and the Marine Corps to establish a standing joint land forces component command in the region. This became Joint Task Force Kuwait (JTF Kuwait) with its headquarters at Camp Doha, Kuwait.* Marines, and Marine units, rotated through the command, whose staff was largely identical with that of ArCent. In the fall of 2001, for example, then-Brigadier General Emerson N. Gardner, Jr., was in command of JTF Kuwait. ArCent and the joint task force morphed into Coalition Forces Land Component Command in late 2001, under the command of a three-star Army general, Lieutenant General Paul T. Mikolashek, who moved his flag to Kuwait in December. Like his predecessor Zinni, General Franks wanted a strong Coalition Forces Land Component Commander, with its own staff, to focus on the ground war while he focused on his own responsibilities for the war at the next higher level. The officer who was to command Coalition Forces Land Component in 2003, Lieutenant General David D. McKiernan, found that there was "huge goodness in that arrangement."**

*In 2003, while serving at Coalition Forces Land Component Commander, Marine BGen Christian M. Cowdrey made much the same point, commenting that the combined/joint land headquarters CFLCC fulfilled a function that would have been difficult for any other organization to fulfill, that of orchestrating the plan. (BGen Christian B. Cowdrey intvw, 26Apr03 (MCHC, Quantico, VA); Tom Clancy, Gen Anthony Zinni, and Tony Koltz, *Battle Ready* [New York, NY: G. P. Putnam's Sons, 2004], p. 315; see also Zinni intvw)

**Kevin Peraino, "Low-Key Leader; LtGen David McKiernan is the Soft-Spoken Soldier with the Hard Job of Commanding U.S. Ground Forces in Iraq," *Newsweek* Web Exclusive, 19 Mar03.

solution. He crafted an arrangement with NavCent for MarCent to come onto the Naval Support Activity itself, even though there was little room to spare on that small base. NavCent agreed to allow MarCent to set up on a baseball field that was, by turns, the source of fine dust and, when it rained, some very sticky mud.[21]

The result was an unusual overlay of temporary structures on a base with some amenities. Some of the MarCent Marines worked in the general-purpose tents on the baseball field. They had some air conditioning, as much for the computers as for the Marines, but were often hot and uncomfortable. Other Marines worked in only slightly more comfortable expeditionary buildings. Like most Marines on major staffs after 11 September, the MarCent Marines worked incredible hours, under great pressure, with little time off. But there was a small upside; less than 100 yards away was a well stocked food court, as well as a gym, exchange, and swimming pool. Sometimes, when the force protection condition was high, one might witness the anomalous sight of half-camouflaged U.S. Navy personnel with loaded rifles sitting or lying among the palm trees and brush outside the food court, on the lookout for any terrorists who might have made it through the perimeter, which was an elaborate affair with concrete barriers, barbed wire, and sentries. This was not quite as farfetched as it sounds; not only was there a continuing terrorist threat, but, especially in early 2003, there were also anti-American demonstrations and occasional terrorist attacks throughout the region. If there had been a firefight, the personnel on base could have watched it while sipping cappuccinos. It was an unusual way to go to war.

MarCent now became the only purely "service component command" in theater, with the enormous responsibility of providing and sustaining thousands, and potentially tens of thousands, of Marines and sailors. Like his counterparts at NavCent; Army Forces, Central Command (ArCent); and Air Forces, Central Command (CentAF), General Hailston reported to the Commander-in-Chief, Central Command, General Tommy Franks, who held sway over all U.S. forces in the region, whatever their purpose. Unlike his counterparts, General Hailston was not a tactician. He would get the Marines to the war, make sure they had what they needed for the fight, and generally offer advice to General Franks on how best to employ them. It was Hailston's job to look 45 to 60 days into the future, anticipate Marine requirements, and then fulfill them, all in all an enormous undertaking. But when Marines went into combat, they

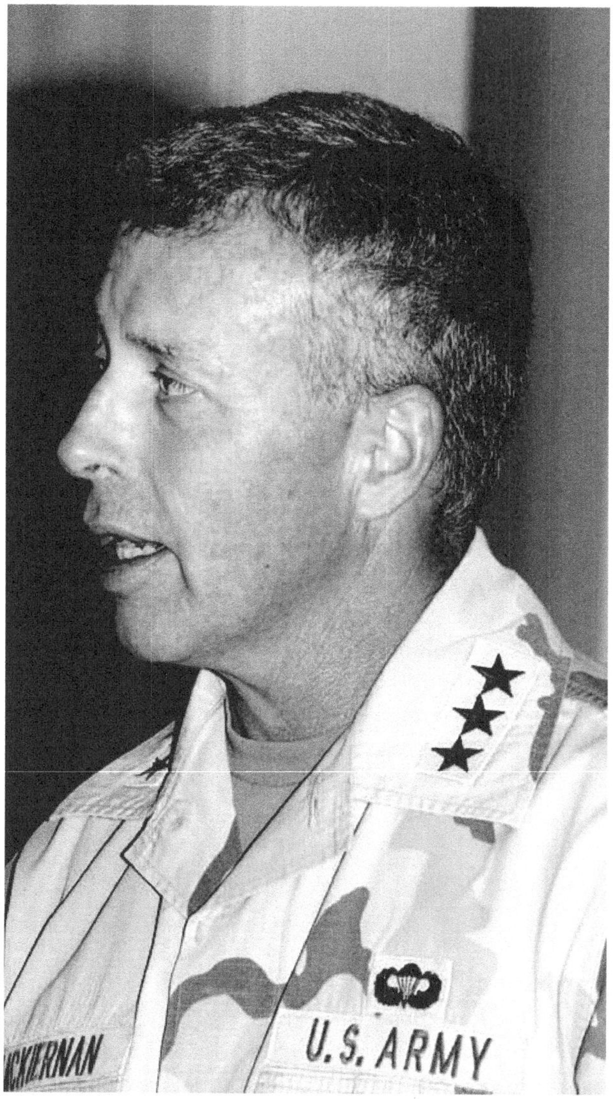

DVIC DM-SD-05-04888

LtGen David D. McKiernan, USA, a graduate of the College of William and Mary, gained experience in the Balkans as a staff officer in the 1990s. In September 2002, he assumed command of the U.S. Third Army and U.S. Army Forces Central Command, and became the Coalition Forces Land Component Commander in preparation for Operation Iraqi Freedom.

would fight under one of the "functional components" like the Coalition Forces Maritime Component Command (CFMCC), which was NavCent's warfighting guise. For example, Admiral Moore became both the NavCent commander and the CFMCC commander. There were similar arrangements for CentAF and ArCent to become, respectively, the Coalition Forces Air Component Command (CFACC) and the Coalition Forces Land Component Command (CFLCC). This was a radical departure from the organization for Desert Storm some 10 years earlier,

Marine Corps Photo 020112-M-1586C-002

Afghans greet Marines from the 26th Marine Expeditionary Unit's Company A, 2d Light Armored Reconnaissance Battalion, on a routine mounted patrol through a village near Kandahar, Afghanistan.

which had been fought without a unified land command.*

This created a dilemma for the Marine Corps, naval infantry and air forces that did not fit neatly into any of these categories. In 2001 the result was another set of complications. Task Force 58 initially reported to NavCent/CFMCC. This was a natural relationship, especially while the task force remained afloat. Once it was ashore, and after CFLCC had stood up in mid-November, it had to develop a relationship with that headquarters, which had assumed responsibility for all operations on land in Afghanistan. Accordingly, on 30 November, CFMCC assigned tactical control of Task Force 58 to CFLCC, without giving up operational control. Especially at first there was some friction between Task Force 58 and CFLCC; the CFLCC culture was "Big Army," the world of large staffs and detailed reporting, certainly a change for the Marines on Task Force 58. But over time everyone made it work, the "TF 58 staff was able to adapt to the new information requirements while continuing to develop a solid working relationship with the CFLCC staff. The positive relationship would last throughout the operation as CFLCC buttressed and represented CTF 58's interests."[22]

*Between 2001 and 2003, the words "combined" and "coalition" were used in the titles of these organizations to mean the same thing. While "joint" refers to U.S. commands with more than one service, "combined" or "Coalition" refers to commands that also include foreign elements. For the sake of simplicity and consistency, I will use coalition.

Most Marine staff officers in the Persian Gulf became well versed in the intricacies of operational control and tactical control. By and large, what General Hailston exercised was operational control, organizing and employing forces, sustaining them, and assigning general tasks, but not tactical control, which was the specific direction and control of forces, especially in combat. The two exceptions were CJTF-CM and an organization known as the Marine Logistics Command (MLC), which would play a prominent role in the Iraq War in 2003. Like the organization for the land war, this, too, was different from Desert Storm. Then the senior Marine in theater, Lieutenant General Walter E. Boomer, had commanded I MEF and MarCent, a heavy burden for one commander, who in both capacities reported directly to General Schwarzkopf.[23]*

General Zinni has argued that especially in a major contingency, each function requires a separate staff, with a different focus.[24] General Hailston's chief of staff, Colonel Stephen W. Baird, believed the arrangement in Desert Storm had stressed the I MEF staff and forced General Boomer to divide his time between warfighting and Service component issues.[25] Having one commander responsible for "shaping, providing, and sustaining" Marine forces and another commander for warfighting would free the warfighter to focus on combat. The Service component commander, who would be senior, could forge a relationship with the commander-in-chief and protect the sanctity of the Marine air-ground task forces, resisting the understandable propensities of non-Marines to break task forces into their building blocks, splitting ground and air components apart and sending them to fight with their counterparts in the Army and the Air Force.[26] (General Hailston agreed with this point of view.) Speaking in late 2003, General James T. Conway added the commonsense point that it would have been difficult for him to command both MarCent and I MEF, since in one capacity he would be the equal of the CFLCC commander, and in the other, his subordinate.[27]

Even with the improved organization of the force, there was some Service rivalry. One notable point of contention was the "force cap" placed on the Marines while Task Force 58 was ashore—CentCom decreed

*Neither Gen Boomer nor Gen Hailston reported directly to Headquarters Marine Corps (HQMC). However, there was a lively exchange of views and data between the CentCom theater and HQMC, especially the office of the Deputy Commandant for Plans, Policies, and Operations (PP&O). During 2002 and 2003, Gen Hailston and LtGen Emil R. Bedard of PP&O were in close contact, discussing and coordinating plans virtually on a daily basis.

that initially there could not be more than 1,000 Marines and sailors in Afghanistan in late 2001. Some Marines interpreted this as a gratuitous slap in the face, "Big Army" making sure the Marines did not steal the show, and, by extension, making it easier for the Army to assume the Marine mission in Afghanistan, a process that was well under way by mid-January 2002.[28] But the Marines complied and were still able to carry on with their mission.

Task Force 58 was released from CFLCC control on 3 February 2002.[29] Although small numbers of Marines stayed in country, either as individuals or as units, for quite some time, the 26th MEU(SOC) left Afghanistan on 13 February. This spelled the end of Task Force 58's engagement in that country. Now the nature of the task force's achievement was even clearer. Apart from validating some of the tenets of expeditionary maneuver warfare, it showed Marines, literally and figuratively, how far they had come from the traditional "two up and one back" mind-set that had characterized the war in Vietnam and, to a certain extent, Desert Storm, which had had some of the characteristics of an old-fashioned linear battlefield.[30]

Task Force 58's experience also showed Marines how far they had come since Desert Storm in another way. Despite the friction over matters like the force cap (some friction was inevitable), there had been nothing like the bad blood and the temper tantrums that are discernable beneath the surface of the memoirs of the Gulf War; in Afghanistan, and then in Iraq, senior participants almost uniformly reported that the level of cooperation was unparalleled, both among the Services as institutions and among their chiefs as individuals. To cite just one example, CFLCC's General McKiernan stated categorically that "there were no rifts" between the various commanders.[31]

There are a number of possible explanations for the relative good feeling, the general atmosphere of trust, among the Services after September 2001. These ranged from the maturing of joint routines at the combatant commands, to the exigencies of the situation, that is, the unifying effect of the events of 11 September, to the backgrounds and the personalities of the commanders. A number of controversial

A Marine crew chief guides a CH-46 Sea Knight of Marine Medium Helicopter Squadron 365 at Kandahar International Airport in Afghanistan. Squadron aircraft flew re-supply, long-range reconnaissance patrol, ground escort, armed interdiction, and heliborne assaults in search of Taliban and Al Qaeda forces.

DVIC DM-SD-02-06319

issues such as the employment of Marine air in the joint environment, which had been a serious point of contention during Desert Shield/Desert Storm, had more or less been resolved during the decade since the Gulf War. CentCom commanders like General Zinni, first and always a Marine but also something of an iconoclast and nothing if not a "joint" visionary, had created and exercised joint structures such as Joint Task Force Kuwait, with its provisions for alternating Marine and Army commanders. Similarly, Operations Northern Watch and Southern Watch, run for years by United States and allied air forces to enforce the no-fly zones in the top and bottom one-thirds of Iraq, were successful combined and joint operations.[32]

Analyzing the backgrounds of the participants in the Iraq War, historians Williamson Murray and Major General Robert Scales, USA (Retired), have made the comment that the officer corps was far better trained and educated in 2003 than it had been during the Gulf War, let alone Vietnam. The key here is the word "educated"; beginning in the 1980s, officers were not only well trained, able to perform battlefield tasks, but they could also think in "operational terms." "The new emphasis," Murray and Scales wrote in their excellent overview of the period, *The Iraq War*, "was on maneuver, deception, exploitation, and decentralized leadership."[33] In the Marine Corps the reform movement had started informally with meetings after hours at Quantico, sometimes over beer, but just as likely over coffee or sodas, and spread little by little to other bases. Many of the ideas that emerged were eventually enshrined in doctrine, particularly *FMFM 1: Warfighting*, during the commandancy of General Alfred M. Gray in the late 1980s. Not everyone became a maneuverist, but it is fair to say that by the end of the 1990s, virtually everyone was familiar with the term and had been influenced by it in some way.[34]

The personalities of the commanders, and how they meshed, were also part of the picture. General McKiernan commented that "the big strength in this campaign was the personalities of the various component commanders. . . . You can say a lot of that [inter-Service cooperation was possible] because of developments in joint doctrine and training . . . but a lot of it [was] . . . also in the chemistry between . . . the leaders."[35] General Franks, with his down-to-earth style, was known as an officer who listened to his subordinates. A strong proponent of military transformation, he was a commander who identified the desired "effects . . . and tasks and purpose, but [left] . . . the planning to the component commanders."[36] He had a long-standing "joint" reputation, having been the ArCent commander under Zinni. It had been a natural progression for him to take the commander's chair and continue the joint tradition. Even after becoming the commander, General Franks had stayed in touch with General Zinni, using him as an unofficial mentor and even trying, in late 2002, to use him as an official mentor for a CentCom exercise. (Zinni did not come to theater, because by then his blunt pronouncements against the coming war had made him *persona non grata* with the Pentagon.)[37] At MarCent, General Hailston had an unusually varied background, having served as an infantryman, an aviator (whose call sign was "Titan"), and a force service support group commander. He had relationships with other senior commanders, especially CFACC's Lieutenant General T. Michael Moseley, USAF, which went back many years. He also had a reputation for not suffering colonels gladly and for taking good care of his younger Marines, not necessarily bad traits to have in 2002 and 2003 (at least so long as you were not a colonel). The field historian assigned to his headquarters, Lieutenant Colonel Jeffrey Acosta, attested to the general's ability to ask penetrating questions across a broad range of subjects, and to keep supplies and equipment flowing to theater.[38]

Murray and Scales assert that General McKiernan, who was to become CFLCC commander in September 2002, "proved to be an inspired choice."[39] An armor officer who was quiet but compelling, McKiernan's background included joint and combined experience, in addition to senior Army commands. He, too, was not afraid of new ideas and wanted to find the best organization for the fight, as opposed to doing things the way they always had been done. He had what *Newsweek* was to call "a temperament as . . . even as the desert," which also made it easy for him to work with other Services.[40]

The officer who set the tone for virtually all Marines in theater was General James T. Conway. A graduate of Southeast Missouri State University, he was commissioned in 1970 and had had a successful career in the infantry. When away from the fleet, he served as the commanding officer of The Basic School and then as president of the Marine Corps University. He also served two tours on the joint staff in Washington. From 2000 to 2002, he was the commanding general of the 1st Marine Division, and then from August to November 2002 he served as the deputy commanding general at MarCent, which gave him an opportunity to work closely with General Hailston on CentCom issues while he waited to take command of I MEF. This was the commandant's ini-

DVIC DM-SD-05-06318

LtGen James T. Conway, commanding general of I Marine Expeditionary Force, addresses the officers of Regimental Combat Team 7 at Camp Coyote, Kuwait. Gen Conway had charge of a battalion landing team during Operations Desert Shield and Desert Storm before assigned command of the 1st Marine Division and then I Marine Expeditionary Force.

tiative. General Jones wanted to maintain continuity on the West Coast and to make General Conway even better prepared for his next job, to command I MEF and to balance its equities against MarCent and CFLCC requirements.[41*]

Murray and Scales described General Conway as "big," he was well over six feet tall, "bluff, well-read, and well-educated," and concluded that he "represented all that was best about the new United States Marine Corps, which General Al Gray as the commandant had built up."[42] He was a popular commander, described as an officer and a gentleman who was good to work for and who took care of his troops.[43] He was nothing if not involved in what his subordinates were doing. For example, he had a policy of wanting to be briefed in person on unusual, high-risk evolutions, as a young British reconnaissance officer was to discover during the war when

*When asked how the right Marines, a virtual dream team, had come to serve in key billets in the CentCom area of operations in 2002 and 2003, the Commandant of the Marine Corps said it had been as much a matter of good long-term personnel policies as of any specific, short-term assignments. (Gen James L. Jones, Jr., intvw, 14Jun04 [MCHC, Quantico, VA])

he was whisked from his position in the desert in order to brief the I MEF commander, in person, on an upcoming operation. It was typical of General Conway to focus on the extraction plan; he wanted to be sure there was a plan to take care of the soldiers and Marines in the worst case.[44] Even months after the war, he remained acutely conscious of the casualties that had occurred during the campaign—able to recite numbers and remember individual cases.[45] He has described his own command style as "democratic," which meant he preferred to command by first listening to his subordinates and then outlining his intent. He knew when to give his subordinates free rein and when to intervene. A review of the journals kept by the field historian at I MEF headquarters during the Iraq War reveals that, like General Hailston, General Conway spent much of the time asking questions and gathering information. They also show that he was typically optimistic, slow to anger, and virtually unflappable, equal to any challenge, whether contemplating the possibility of a chemical attack or dealing with a difficult counterpart or subordinate, traits that he would need in Operation Iraqi Freedom.[46]

Chapter 2

Inside Our Own Loop: Joint Planning for War in Iraq

Operation Enduring Freedom set the stage for Operation Iraqi Freedom. The second operation was not a clear-cut sequel to the first; for military planners, there was no straight line from success in Afghanistan in the winter of 2001-2002 to a war in Iraq in 2003 to remove the dictator Saddam Hussein from power. But Enduring Freedom was in many ways the starting point for Iraqi Freedom.

Coalition Forces Land Component Command (CFLCC) and Coalition Forces Maritime Component Command (CFMCC) were still very much in existence when the focus shifted from Afghanistan to Iraq early in 2002. The Marines in Task Force 58 who fought in Afghanistan between November 2001 and January 2002 had operated under both of these commands and had developed good relationships with them. If anything, General Franks was more convinced of the need for an organization like CFLCC in a war with Iraq. In Afghanistan there were few U.S. troops on the ground, but a war with Iraq could be a larger, more complex undertaking by far and most likely would be won or lost on the ground. Alongside CFLCC and CFMCC, Coalition Forces Air Component Command and Coalition Forces Special Operations Command remained very active commands that the Marines would engage.

Both for the individuals who went to Afghanistan and for the organizations that sent them, the experiences of Task Force 58 in Enduring Freedom set some of the specific conditions for war in Iraq. Integrated into a combined, joint operation that fused airpower, special operations, and information operations, the Marines had operated hundreds of miles from the beachhead, relying heavily on Marine airlift, especially by Sikorsky CH-53E Sea Stallion helicopters followed by Lockheed KC-130 Hercules cargo carriers. They had succeeded without a large staff, or a plan that was hundreds of pages long, relying instead on common sense, good liaison officers, and "hand con" (not a formal relationship like tactical control or operational control but one sealed with a handshake).

The focus began to shift to Iraq even before Task Force 58 left Afghanistan. In the wake of 11 September, the administration had looked to see if Iraq was behind the attacks on the World Trade Center and the Pentagon before deciding to fight in Afghanistan. But it seems that Iraq was never far from the administration's mind and that while overthrowing the Taliban and uprooting Al Qaeda were short-term objectives, Iraq, more specifically, removing the regime of Saddam Hussein, had always been a long-term objective. Military planners followed the administration's lead on both Iraq and Afghanistan.[47] In the fall of 2001, staff officers from Headquarters Marine Corps to Central Command (CentCom) to CFLCC were considering the possibility that U.S. forces could be called upon to invade Iraq. There were some preliminary planning directives, but even without them many Marines and Army officers simply assumed that Iraq would come after Afghanistan. CFLCC's Major General Henry W. Stratman, USA, spoke for many when he said that after 11 September the assumption was not whether, but when, the United States would go to war with Iraq.[48]

In January 2002, General Hailston, in his capacity as the Commanding General, Marine Forces Pacific, directed I MEF to focus its efforts on preparing for "contingencies" in the CentCom theater. "CG, MarForPac . . . decided to focus I MEF efforts on preparation for contingencies in the CentCom theater. I MEF's role in PacCom activities was minimized or assumed by III MEF and MarForPac" to the virtual exclusion of other activities.[49]

This was when I MEF's majors and lieutenant colonels, along with a few colonels, who make any large staff run earned their pay. They entered into what was for many of them the most intense period of their careers in the Marine Corps, one that would not let up for some 18 months. Even before they deployed from the United States, they came close to spending every waking minute working on the plan, often in windowless secure spaces. When they were not working in a vault, they might be traveling from one drab base to another for a conference or a war game. They no longer had any time for themselves, let alone their families or their "honey do" lists.

For the I MEF intelligence section, the focus on CentCom meant embarking on "a wide variety of activities, including presentation of many staff orientation and mission analysis briefings, . . . supporting estimates and plans, . . . development [of require-

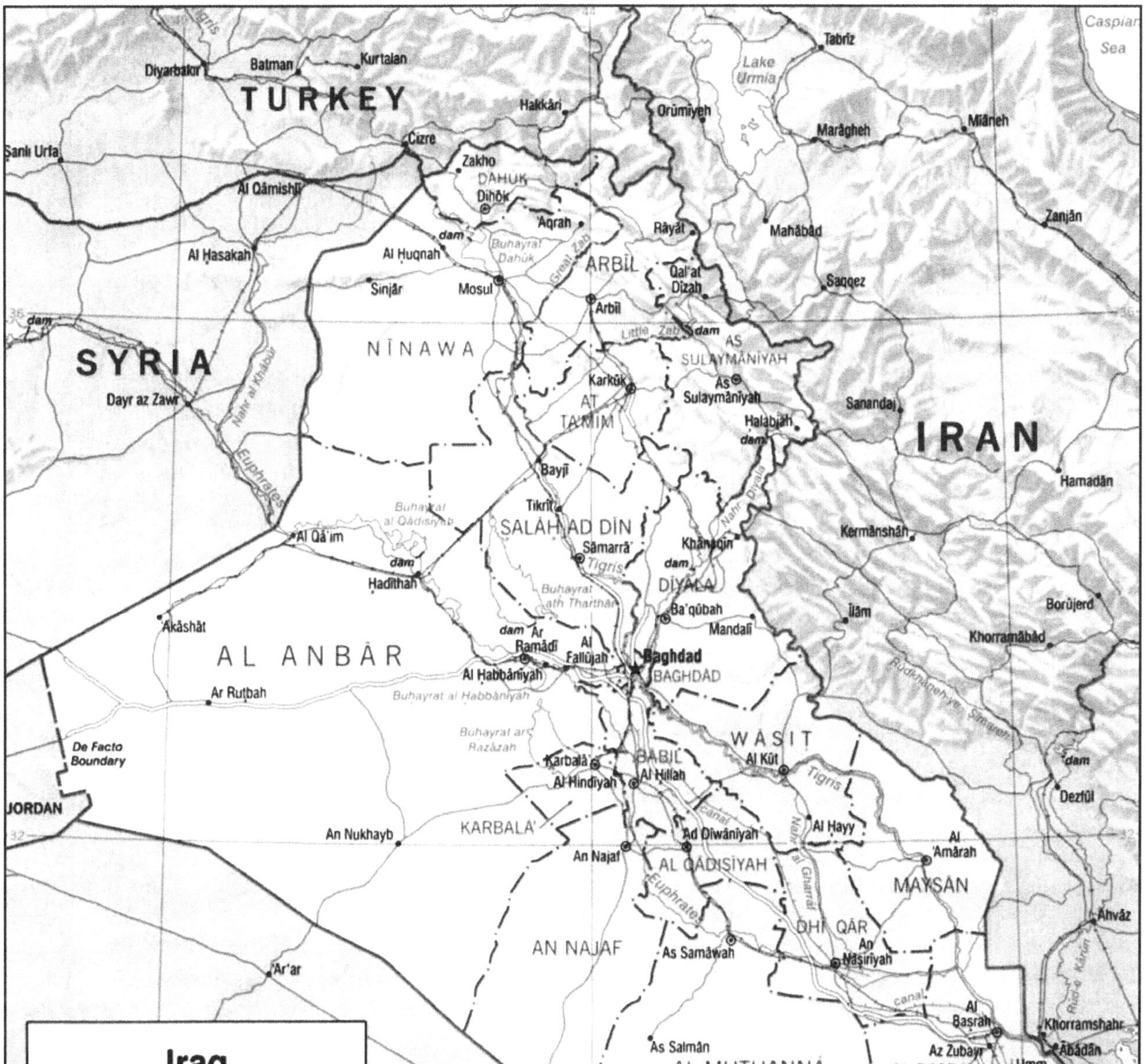

Base 802669AI (R00667) 12-99

ments] . . . hosting visits from national and theater intelligence organizations (CIA, DIA, MCIA, CentCom, ArCent, and V Corps) and [making] liaison trips." For its part, I MEF's current operations section became involved in various exercises in the CentCom area of operations. In April, for example, it participated in the exercise "Lucky Sentinel," a combined/joint computer-assisted command post exercise designed to train and sustain the battle staff of Joint Task Force Kuwait. It was conducted "in conjunction with ArCent; CentAF; and the Kuwaiti military," good practice for the events that were about to unfold.

Similarly, I MEF's future operations section used exercises like "Desert Scimitar" and "Lucky Sentinel" to prepare for war in the Middle East, while the MEF plans section was involved in longer-range, high-level operational planning. In the subdued words of the I MEF command chronology, "G-5 directed most of its efforts . . . [to] . . . details [of] I MEF's slice of the USCincCent's plan in concert with the nation's strategic objectives."[50]

This was a dramatic understatement. For a few months, the plans section took the lead in the intense and exhausting task of laying the groundwork for Marines to participate in a war for Iraq. In January 2002, General Hagee sent one of his lead planners for Korea, Lieutenant Colonel George W. Smith, Jr., to Tampa with Colonel Jonathan G. Miclot, the plans officer at 3d Marine Aircraft Wing. Their mission was to represent MarCent, not just I MEF, on CentCom's long-range planning element.[51] This was a happy consequence of Marine staffing practices. In his MarCent capacity as a component commander, General Hailston made the decision to let I MEF, the warfighting command subordinate to him, play the leading role in operational planning.

The long-range planning element was small and run mostly by Army ground officers, who had been working in the same directorate as Major General Keith J. Stalder, an even-tempered Marine aviator who was the deputy J-3 at CentCom and would become the deputy commanding general of I MEF later in the year. Since late 2001, the focus of the planning element's much compartmented work had been Iraq, and the timeline was short—this was not theoretical planning for some unlikely contingency in the distant future. The word was that CentCom might need to be ready to fight as early as the spring of 2002; this could be a "come as you are" war. In that regard it would not be unlike the campaign in Afghanistan, which had been a relatively quick success.

Commissioned in 1968 through the Enlisted Commissioning Program, LtGen Earl B. Hailston went to flight school and served in a variety of aviation and ground assignments before assuming the multi-hatted position of Commander, U.S. Marine Forces Pacific/Commander, U.S. Marine Forces Central Command/Commanding General, Fleet Marine Force, Pacific/Commander. U.S. Marine Corps Bases, Pacific, on 10 August 2001.

Photo courtesy of Field History Branch

The vision that guided the planning was to win by creating "shock and awe" through multiple lines of operation putting simultaneous pressure on the enemy—from the air, from conventional ground operations, and from various kinds of special operations, to include "non-kinetic" operations and operations by proxies like the Kurds. There were three main groups in Iraq—the Shia majority, the ruling Sunni minority, and the Kurds. The Kurds lived a more or less autonomous existence in the northeast corner of the country and had large, well-armed militias. Neither the Kurds nor the Shia had much love for Saddam Hussein, who had suppressed them in unimaginably brutal ways. For Marines, "shock and awe" was something like the "combined arms effect," on a grand scale, of forcing the enemy into a series of dilemmas he could not resolve; if he turned to face one threat, he would make himself vulnerable to another threat. It was something like facing mortars and machine guns at the same time; was the infantry better off staying in fixed positions during a mortar attack, or getting out of its holes into a field of machine gun fire?

While often associated with the air offensive in what was not yet officially known as Operation Iraqi Freedom, "shock and awe" was more than a theory of air warfare. The concept has both a recent and a more distant past. In the recent past it can be traced to a book published by the National Defense University in Washington, D.C., in December 1996 by Harlan K. Ullman and James P. Wade titled *Shock and Awe: Achieving Rapid Dominance*, with contributions by retired Generals Charles A. Horner, USAF, and Frederick M. Franks, Jr., USA, both of Desert Shield/Desert Storm fame, and retired Admiral Leon A. Edney, who had been commander-in-chief at Atlantic Command. The authors' purpose was to offer an alternative to the strategy of overwhelming force—sometimes called the Powell Doctrine, on display in Desert Storm, by pointing to the potential of the many new technologies to achieve "rapid dominance" that would paralyze the enemy's will to resist, ideally but not necessarily before any ground forces were committed. It is easy to see why this doctrine is especially attractive to the Air Force, as it suggests that airpower alone could be decisive.

The authors of *Shock and Awe* readily conceded that their theory was not entirely new, with antecedents in the World War II concept of "blitzkrieg" and various operations since. Iraq war historians Williamson Murray and Major General Robert Scales discuss how General Anthony Zinni and his Army counterparts used ideas like "rapid dominance" and "overmatching power" when they considered joint contingency plans against Saddam Hussein.[52] Zinni himself has commented that after Desert Storm, a more or less traditional war, he was convinced that the Marine Corps needed to learn to think along multiple lines of operation. Marines would also have to work better in the joint arena. These were, he said, the lessons he tried to inculcate in I MEF after he became its commanding general in 1994. He added the sage comment that, like many, he imagined at the time that he was on the cutting edge, but realized later that the winds of change were blowing in other places in the Marine Corps around the same time. Transformation is not a straightforward, top-down process.[53]

Looking back to 1989, Murray and Scales found an interesting precedent, a small war before Desert Storm that was almost like a laboratory experiment of the ideas that dominated planning in 2001 and 2002. This was Operation Just Cause in Panama: "Maneuver in Panama was nonlinear and focused on control of the whole operational area rather than on the sequential capture of key terrain and high ground characteristic of more traditional forms of maneuver."[54] Just Cause was complemented by new forms of technology such as laser-guided bombs that enabled pinpoint targeting, that were to improve markedly over the next 10 years, in turn enabling further strides in doctrine. Murray and Scales concluded that Just Cause had little effect on Desert Storm, which did not incorporate much of this kind of "shock and awe," and added that the U.S. military's lack of preparation for the postwar period in Panama had contributed to widespread looting and lawlessness after the fighting had stopped.[55]

Throughout 2002 and into 2003, the basic concept for a war of "shock and awe" against Saddam Hussein did not change. General Franks made sure of that. Nor did other threshold concepts change once they had been established. These had to do with the basic organization for combat, how the Marines would organize for the fight, who they would report to, and with basing the 3d Marine Aircraft Wing in theater. Smith and Miclot found that their counterparts in the long-range planning element were working from a scenario that had two Marine expeditionary brigades performing various missions, mostly to do with security, in southeast Iraq while the Army's V Corps carried the fight to the enemy in the north. Over the next 40 days, the two Marine planners worked patiently to lay down the Marine expeditionary force "marker"; the argument that the Marines should fight as one expeditionary force in Iraq, the whole force being greater than the sum of its parts, let alone two independent Marine expeditionary brigades.[56]

This took some doing. Although it was something they had always known and heard, Smith and Miclot learned again, firsthand, just how peculiar the concept of the Marine air-ground task force is to non-Marines; no other Service has anything quite like it. Many Marine and Army units of apparently comparable size are not in fact comparable; the Marine unit typically has more organic power, because it comes with its own air support. This is one of the factors that led to disconnects when joint planners were placing Army and Marine units on the board. As Lieutenant Colonel Smith put it, it was difficult to get into the Army's "comfort zone," to make his Army counterparts comfortable with the "MEF single battle" concept, but he felt that after 40 days of hard work, he and Miclot had succeeded.[57]

The other threshold issue they took on was "beddown" for the Marine aircraft wing, essentially a matter of forward basing. This may not sound like a particularly dramatic issue, but with the U.S. Air Force occupying ever more space on the air bases in Kuwait, it was important for the Marine Corps to stake claims to space for its aircraft near the front. Otherwise the wing would have had to look for bases farther afield, which would degrade its ability to get into the fight and especially to provide responsive close air support. Miclot worked hard and succeeded; Smith considers him one of the unsung heroes of the war for identifying, and resolving, the issue early on.[58]

On 12 February 2002, in Washington, Smith and Miclot back-briefed the trio of officers who held the key positions for shaping basic Marine Corps policy and "major muscle movements" in 2002 and 2003—Lieutenant Generals Hailston (in his MarCent capacity), Hagee (in his I MEF capacity), and Emil R. Bedard (the deputy commandant of the Marine Corps for Plans, Policies, and Operations (PP&O), basically

Photo courtesy of Col Charles J. Quilter II

An Iowa native, Col Jonathan G. Miclot was commissioned from the U.S. Naval Academy and then designated a naval flight officer in 1981. He commanded VMFA(AW)-225 before being assigned as plans officer, 3d Marine Aircraft Wing.

the commandant's current operations division). Smith and Miclot were gratified to find that the generals wholeheartedly supported their work on both the unity of the Marine air-ground task force and the bed-down issues, something that did not change for the life of the operation. Within a few months, General Hagee would become commandant and General Conway would take his place at I MEF; but at the top of the Marine Corps, the I MEF-MarCent-PP&O nexus remained the central forum for consultations and decisions about Iraq and certainly played a prominent role in the sourcing conference that took place later in February to come up with a preliminary troop list for Iraq. (The levels of involvement shifted somewhat over time. Initially, Bedard played the most prominent role. As the lay-down for Iraq started to gel, MarCent and I MEF played more prominent roles, while Bedard tended to monitor developments. Although under CentCom's operational control, MarCent also reported to Headquarters Marine Corps, generally to Bedard, on strictly Marine Corps business such as deciding which Marine units to deploy to CentCom or how to outfit them.)[59] Thanks to the state of technology for secure communications, the commanders and the planners were able to stay in very close touch throughout the process, and the senior officers who were read in on Iraq could develop and maintain arguably the best situational awareness in the history of warfare. They could find out almost anything they wanted to know.[60]*

It was Marine Corps doctrine that the Marine expeditionary force should not only fight as a Marine air-ground task force, but that it should also plan as an air-ground task force. Planners for the constituent parts of the task force should integrate their work. They should not work as stovepipes, waiting to interface at senior levels after plans were well advanced. Instead they needed to function as a network, at all levels, from the start.**

Officers of I MEF used the same approach in their work with other Services. In the interests of coordinated planning, Colonel Joseph D. Dowdy, and Lieutenant Colonel Smith, who was the I MEF plans officer at the time, reached out to their counterparts during the many planning conferences that took place over the next few months. These included sessions at Transportation Command at Scott Air Force Base in Illinois; V Corps headquarters at Heidelberg, Germany; and CFLCC headquarters at Camp Doha, Kuwait, not to mention the commanders' conferences chaired by CentCom on a regular basis. The Marines were generally able to establish and maintain good relationships with their counterparts, especially at the working level. Sometimes it even reached the point where planners identified more with one another than with their parent commands, sure sign that relationships had gelled.[61]

It was always assumed that I MEF would fight under CFLCC. When CFLCC had taken control of land operations in Afghanistan in November 2001, CentCom had charged it with the traditional Joint Task Force Kuwait missions of defending Kuwait and generally being prepared for war with Iraq. That had not changed in 2002, and from the start CFLCC had played the central role in planning for the ground war against Iraq and for I MEF's role in it.[62] The arrangement was that, exercising operational control, MarCent would flow I MEF to theater and provide for its sustainment, relying mostly on the Marine Logistics Command drawn largely from the 2d Force Serv-

*While in Tampa, LtCol Smith was able to hold a secure telephone conversation with Gen Hagee virtually every day, briefing him on developments and receiving his guidance.

**This was one of the fundamentals of the Marine Corps planning process: "continuous planning requires continuous coordination laterally and between echelons as plans are adjusted and refined over time." (U.S. Marine Corps, *Planning* [Washington, D.C.: Department of Navy, 1997], p. 83) ice Support Group for that purpose. But then MarCent would assign I MEF to CFLCC's tactical control for combat operations alongside the Army's V Corps. The result, General McKiernan commented later, would be the first time since the Korean War that there would be a combined "operational-level, land component command/warfighting headquarters."[63]

Like planning, educating other Services about the Marine air-ground task force and safeguarding its equities was a continuous process. It went on long after the initial lessons in Tampa in January and February 2002. Marines at many levels engaged their counterparts at CFLCC and other commands, finding ways to make the lessons stick without being resented by their "students." The process started at the top. From the highest levels on down, Marine commanders and planners stayed on message. General Hailston continued to defend I MEF's identity as an air-ground task force. While still I MEF commander, General Hagee did the same when meeting with General Mikolashek, who remained in command at CFLCC through the summer of 2002. Then, when General Conway was preparing to replace General Hagee as I MEF commander and General McKiernan became the new general on this particular block, Conway hosted McKiernan and his subordinates at Camp Pendleton. First the Marines presented the I MEF capabilities brief. Among other "lessons" about the air-ground task force, General Conway wanted to make sure his new boss understood that "our air" was also "his air," a concept General McKiernan came to embrace.[64] He also wanted to give his subordinates—now-Major General Mattis, who had become commanding general of the 1st Marine Division, and Major General James F. Amos, the 3d Marine Aircraft Wing commander, a chance to talk through issues with CFLCC planners. The specific issues were perhaps not as important as opening channels of communication among general officers at CFLCC and I MEF; the participants remember feeling that the meeting cleared the air. General McKiernan's view, that CFLCC was there to "shape" the fight by its subordinates, not to plan it, must have gone over well with the Marines.[65]

The generals set the tone for their respective commands, and much the same process happened at lower levels as subordinates worked their way through the many practical issues involved in joint operations. General McKiernan characterized the prevailing attitude throughout CentCom as: "Let us coordinate, and let us cross talk, and then come . . . together at a series of . . . conferences. . . . I would say that that was always done very well."[66] In addition to the exchanges between counterparts, most commands also made a point of exchanging competent liaison officers, just as General Mattis had done in Afghanistan. One of the important lessons learned in Afghanistan and Iraq seemed to be that commands understood the importance of finding strong officers to serve as liaison officers; liaison officer was no longer a suitable billet for an underachiever whom a commander wanted out of sight and mind.[67]

Technology helped the process along. Counterparts and liaison officers may have held personal meetings whenever they could, but they also could and did look each other in the eye almost every day over secure video-teleconferencing links—this happened before, during, and after combat. Virtually all of the generals in theater were regular and, it seemed, enthusiastic users of this technology. At the same time, there was a robust exchange of e-mails and discussions in chat rooms on the SIPRnet (the secure military internet system), where officers could also consult the drafts of one another's plans and work through revisions. In short, there were unparalleled opportunities for the secure coordination of operations, especially in peacetime; operational security did not have to be the obstacle to efficient planning it had been in virtually every other major war before Operation Iraqi Freedom. It was no longer as true that senior headquarters imposed operation plans and orders on subordinate commands. Looking at the process of planning for Iraq, a Fort Leavenworth study concluded that since networks in "the information-age . . . enable, . . . distributed, parallel planning, V Corps, I MEF, and the subordinate divisions were near-equal architects for the final plan."[68] This conclusion assumes that the subordinate commands had the requisite clearances to access highly classified files.

In retrospect, integrating with CFLCC appears to have been relatively easy compared to working with CFACC. Groups of Army officers, especially at CFLCC and in parts of V Corps, came to accept and even embrace the concept of the Marine air-ground task force as an organization with integrated ground, air, and support assets. Although there were nuances in the picture that make it difficult to generalize, it is safe to say that the same was never true to the same degree of CFACC, the Coalition air forces commanded by Lieutenant General Moseley, who was not only a functional and a component commander, but also one whose forces had been conducting operations in theater for quite some time, especially Operation Southern Watch. They were flying combat missions under his command while his counterparts

Photo courtesy of Field History Branch

Gen Michael W. Hagee, Commandant of the Marine Corps, receives a brief on current operations from LtGen Earl B. Hailston, Commanding General, Marine Forces, Central Command.

were still thinking about what they would bring to the fight if and when it occurred.

Before leaving CentCom, General Zinni had laid some of the groundwork for cooperation between CFACC and I MEF by instituting a standard operating procedure for joint fires, which addressed "battlefield coordination, direction, and procedures for . . . air and ground-based fires systems" and was ratified by all of the Service chiefs.[69] He had his Army and Marine Corps subordinates work out an arrangement for the employment of Marine air under Joint Task Force Kuwait, and established the general principle that Marine air would support Marine ground forces, offering any "excess" sorties to CFACC.[70] In 2002, while he was still at I MEF, General Hagee and General Moseley renewed the same general agreement about the employment of Marine air with General Moseley:

> Lieutenant General Moseley . . . , Lieutenant General Hagee . . . , and Lieutenant General Hailston . . . met in June . . . and agreed to make CFACC the air space control authority, with I MEF MACCS [Marine Air Command and Control System] controlling air in support of I MEF. I MEF would publish its own direct support air tasking order (DSATO) to task I MEF aircraft, which was to be included in CFACC's theater ATO [air tasking order]. When Lieutenant General Conway took command of I MEF from Lieutenant General Hagee, the arrangements and relationships did not change; Lieutenant General Moseley continued to endorse the principle of I MEF MACCS controlling air assets supporting the MAGTF [on the] ground.[71]

When interviewed in the spring of 2003 about cooperation with CFACC, Generals Hailston and Amos were upbeat about the subject. They reported general agreement with General Moseley on the role of Marine air in the looming conflict, one that was different from that in Desert Shield/Desert Storm when a fair chunk of Marine air had been split off from I MEF and worked for the equivalent of the CFACC.[72] As General Amos put it, he found General Moseley to be a commander who readily understood the utility of the Marine aircraft wing as a part of the Marine air-ground task force while asserting his own rights to the air space over the battlefield.[73]

The problem, once again, was finding a way to get into the joint comfort zone—to get things right at the working level, to focus not on general agreements but on specific details that would apply in 2003. The default setting at CFACC was to control all of the air space in the area of operations. The Air Force liked to control air space through the air tasking order, described as "the daily master plan . . . [which] listed all of the strikes, CAPs [combat air patrols], tanker missions, and other supporting functions for a 24-hour period."[74] Air tasking orders were prepared about 96 hours before their time of execution. Accustomed to decentralized planning, and interested primarily in supporting I MEF's scheme of maneuver, the Marine aviators had not changed overmuch from Desert Shield/Desert Storm, when they had been "deeply suspicious" of what they saw as an inflexible system that might not be able to respond to last-minute requests for support, that is, the Air Force system seemed to be better suited for strategic or operational offensives than for the kind of tactical uses that were the bread and butter of Marine air. In the end, in this new war the agreement among the generals was to "nest" Marine command and control under CFACC. There would be a Marine air tasking order within the CFACC tasking order; the primary mission for Marine air would be to support the I MEF scheme of maneuver; excess sorties would be made available to CFACC, and there would be provisions for the reverse to occur as well. Excess CFACC sorties often "volunteered" to fly Marine missions; they seemed to enjoy working within the Marine air control system.[75]

A related and very complex issue was the separation of what came to be known as "Air Day," the day the air offensive would begin, and "Ground Day," the day the ground offensive would begin. These terms were commonly abbreviated as "A-Day" and "G-Day." (In this war, there was an often-confusing mix of civilian and military acronyms whose meanings were not entirely clear to everyone.) To summarize what was a lengthy and sometimes hard-fought set of transactions, CFACC thought the war should start with its house brand of "shock and awe," attacking carefully selected targets with precision munitions, some simultaneously, others in a particular sequence. The main targets were the air defense capabilities that had survived Operations Northern and Southern Watch, as well as leadership and command-and-control nodes. With their coverage of nearly two-thirds of Iraq, Northern and Southern Watch represented an early, ongoing, and effective suppression of enemy air defense campaign.

At times it seemed that CFACC thought the air offensive could win the war by itself; there was a precedent of sorts in Serbia when NATO air attacks in 1999 had, by themselves, brought about the desired result: "in every way that mattered, airpower won the fighting in Kosovo, while ground units served to consolidate that victory."[76] This fit with the view, held by many in Washington, that Saddam's regime was held together largely by the threat of force, and for that reason his support was weak and shallow. This was a variant on the "kick in the door and the house will collapse" train of thought, assiduously promoted by people like Iraqi exile Ahmed Chalabi, that seemed to guide a lot of the planning for this war.[77]

It followed, for those who held this belief, that CFACC planning did not have to be as integrated as planning by other CentCom components and that CentCom should allow CFACC enough time to create its war-winning effects. That would be something on the order of 30 days, not too different from the 38 days in Desert Storm that had preceded the ground war; in that war, the two campaigns had not been integrated but sequential. Now, in 2002, CFACC wanted a boundary that put Baghdad under its control for those 30 days, as well as the use of Marine fixed-wing assets that, presumably, the Marine Corps would not need until the ground war started. This was not only an argument against synchronicity; it also undermined I MEF "single battle" doctrine. Simply put, the Marines did not want to break up the air-ground task force, even temporarily.

Unwilling to give up the benefits of synchronizing the air and ground campaigns, many Army and Marine officers consistently argued for a much shorter air offensive. Lieutenant Colonel Smith remembered that as early as late February 2002, the prevailing view at CentCom was that the offensive should last no more than 48 hours.[78] Within I MEF, General Amos continuously repeated that his first and most important priority was supporting the Marines on the ground. General Mattis was one of the leading proponents of synchronicity, making his arguments forcefully throughout the planning phase. He wanted almost no preliminary air attacks before the ground attack and for the air and ground offensives to start very close to the same time. One of the planners at division, Lieutenant Colonel Paul J. Kennedy, came to the conclusion that it was Mattis who won I MEF over to his way of thinking by "socializing" (that is, effectively promoting) this concept, which stood the CFACC concept on its head. If the air and ground campaigns were synchronous, or nearly synchronous, the Marine aircraft wing would have few airframes to spare for CFACC, because it would be busy supporting I MEF.[79]*

No matter who originated the idea, I MEF consistently argued for a much briefer air offensive in its dealings with CFLCC, which adopted much the same position, and made its arguments to other elements in the CentCom chain. These arguments became more compelling in late 2002 as CentCom focused more and more on the southern oil fields after concern for their preservation turned into a strategic imperative. In the end, one of the most telling arguments against a lengthy preliminary air offensive was that it would put Saddam Hussein on notice that the ground offensive was coming and would give him time to sabotage his own oil fields, as he had in the Kuwaiti oil fields during the Desert Storm air offensive. It was largely for that reason that in December 2002 and January 2003, I MEF and CFLCC joined forces to argue for no preliminary shaping whatsoever, which did not resonate with CFACC. The final prewar consensus was that the air offensive should be relatively short and that the two phases should be as closely integrated as possible.[80]

The commander of the British air component in Iraq, Air Vice Marshall Glenn L. Torpy offered a good summary of the factors at play in the final stages of the debate in early 2003, which suggests how difficult it was to close the gap between CFACC and the other components:

> As we developed our thinking . . . there was a shortening of that phase [the air phase] and it came down in the early part of . . . [2003] from approximately 16 days . . . to a matter of five days. . . . [T]hat was driven even closer together, as we got closer to the likelihood of

*LtCol Kennedy was right in so far as Gen Mattis was a very effective advocate for his ideas. However, it appears that others at I MEF and CentCom had reached the same conclusion on their own. (LtCol Paul J. Kennedy intvw, 6Nov03 [MCHC, Quantico, VA])

> the operation ['s] being executed, for three factors. . . . First of all, there was a growing realization that we needed to secure the southern oil fields as swiftly as possible to prevent any subsequent damage. . . . There was nervousness by the American land component and by General Franks over the vulnerability of having a very large land contingent in a fairly small area in Kuwait [waiting for the air campaign to end]. . . . General Franks felt that if he had the ability to synchronize the components together as comprehensively as possible then he would have the [best] chance of . . . getting the campaign over and done with as quickly as possible.[81]

In the days just before the war, the commanders appear to have agreed to shorten the air campaign even further. According to the Fort Leavenworth study, the final plan was for the gap between A-Day and G-Day to be 15 hours. This happened after "Colonel Kevin Benson, the CFLCC C-5 [with whom Marine planners had an excellent relationship] developed and forwarded to the CentCom staff a series of position papers advocating adjust [ments in] . . . the G-A Day sequence."[82*]

The dispute over the separation of A-Day and G-Day went hand in hand with a painful set of disputes over the time-phased force and deployment data (commonly known as TPFDD, closely related to the TPFDL, with the "L" standing for "List"), the computerized system for getting U.S. forces to a fight in good order. It could phase forces to match a plan, and make sure the support they needed would arrive at the right time. Especially in a small, single-Service contingency, this was the kind of rational process everyone was comfortable with. In large deployments the process was trickier; there was a finite amount of lift, especially airlift that the Services had to share. Most of these assets belong to Transportation Command. When the debate over the sequence of the campaign has not been settled, and when no one knows when the war will start, the result can be a three- or four-sided scramble for scarce resources. This is generally what happened between January and July 2002. No one had enough forces in theater at that point; there was not even a firm date by which

Photo Courtesy of Col George W. Smith, Jr.

A graduate of the University of North Carolina, LtCol George W. Smith, Jr., was commissioned in 1985. Following several staff and school assignments, he reported to headquarters I Marine Expeditionary Force in July 2000 where he served consecutively as a future operations planner and a regional plans officer.

everyone had to be ready to cross the line of departure. But there was strong and continuing pressure to be ready to flow forces to theater, which lent some urgency to the discussions about the deployment data. Believing they would start the fight, and having their own plan for that phase (which they had developed more or less on their own), CFACC planners argued that they should be first in line. But if CFACC won the deployment data argument, then it mattered less who won the argument about the separation of A-Day and G-Day, because the ground forces could not be in theater and ready to fight until CFACC had finished moving its forces. In other words, the danger was that the deployment data could drive the war plan, which was the reverse of what was supposed to happen, the deployment data was designed to be a tool for planners. We had, Lieutenant Colonel Smith concluded wryly, gotten inside our own "observation/orientation/decision/action" (OODA) loop.[83*]

There were additional complications to do with

---

*On 1 March, Gen James Conway commented that the air campaign was likely to be brief in order to achieve surprise, which implied a very short separation between A-Day and G-Day. This was consistent with the scenario for the 10 March 2003 I MEF rehearsal of concept drill, when General Conway reminded his staff not to "expect a return of A and G separated by multiple days."

*The OODA loop is a concept pioneered by U.S. Air Force Col John Boyd. Boyd's argument was that if you want to win an aerial dogfight, you will go through this loop faster than your enemy, that is you will get inside his OODA loop.

the relative priorities for moving Army and Marine Corps assets, with the Army generally wanting to lift more than the Marine Corps. Various forms of these disputes continued for some time, leading one anonymous "wag" to comment: "[W]ar is simply the continuation of service politics by other means." In the end the core dispute was not resolved as much as sidestepped. In the diplomatic words of Fort Leavenworth's detailed look at the Army in Operation Iraqi Freedom, the deployment data "lacked the flexibility and responsiveness required by senior leaders," which meant the Pentagon decided not to use it.[84] In early 2003, CentCom would use another process to flow troops to theater. Whatever CentCom or the Pentagon decided, the Marine Corps always had the advantage of being able to fall back on its habitual relationship with the Navy and the maritime prepositioning ships, preloaded with heavy equipment, in order to get to a fight, not to mention the fact that the Marine air-ground task force was by its very nature designed for deployment even before the first planner sharpened the first pencil.

Yet another important part of the planning process was intertwined with the disputes about the timing of the offensive and the use of the deployment data. This was perhaps the most important piece of the interlocking puzzle for the Marines, the base plan for the ground war. Throughout 2002 the plan went through a number of major and minor changes, approximately five of the former and some two dozen of the latter.[85] There are a number of complementary explanations for this. One is relatively simple, so simple that it seems almost trivial. It falls into the "for want of a nail" school of historical writing, meaning that some small details can matter a great deal. It was Benjamin Franklin who said, "for want of a nail, the shoe was lost; for want of a shoe, the horse was lost; for want of a horse, the rider was lost." This explanation has to do with the U.S. military's near-obsession with "bulletized" PowerPoint briefings, which are easy to prepare on a computer and then transmit over the internet. The "bullets" in a briefing may summarize months of careful work, or they may stand alone with nothing to back them up. What often happened in 2002 was that one PowerPoint briefing led to another, a planner would come up with a carefully reasoned course of action and summarize it in a brief. Then a commander, perhaps two or three echelons higher, or even in another Service, would ask a question or order a change. The result could be a course of action that looked good on a PowerPoint slide but had not been thoroughly staffed. That part of the plan would be "one PowerPoint brief" deep, and would either collapse of its own weight or have to be rescued by planners scrambling to do the staff work to back up the change.[86]*

The Marines often thought the other Services pressed for too much detail too early, which they resisted on the grounds that once something was on a PowerPoint slide it could look more final than it was. The bottom line is that the SIPRnet was sometimes a double-edged sword. It usually made concurrent planning easier, but it also made it easier to circulate half-formed ideas, which could, and did, lead to extra work for planners who were already so busy that they were close to forgetting the names of their children.

Another explanation for the nature of the process is that Secretary Rumsfeld had a particular vision he wanted to implement. He was nothing if not an advocate of transformation, and he often made it clear that the military establishment was moving too slowly in the direction that he, and many military thinkers, wanted to go. He certainly did not want to approve a plan anything like Desert Storm. There was simply no need for it, especially after the Afghan operation had demonstrated the potential for the new way of war, with its innovative and very joint lines of operation. There was also the use of "smart" munitions that were both efficient and effective; five bombs from one aircraft could now achieve the same effects that all the bombs from five aircraft had tried to achieve in 1991. Complementing the transformational argument was the military-political argument identified with the neoconservative movement, which had a long-standing policy on Iraq. Bolstered by inside information from Iraqi exiles around Ahmed Chalabi, senior Pentagon officials focused on the fact that the Iraqi Army was a shadow of its former self; it was about one-third the size it had been when Saddam invaded Kuwait and would crumble under an American assault. In addition, the Iraqi people were dissatisfied with the regime, and the majority of them would welcome the invaders as liberators. After the invasion, the United States and its Coalition partners could soon draw their forces down, and the Iraqi opposition could step in to run the country. The bottom line for the Pentagon was that the United States did

---

*Maj Evan A. Huelfer, USA, was CFLCC's lead planner and a great source for historical data. He and LtCol Smith, I MEF's lead planner, used almost identical wording in discussing this phenomenon. Similar ideas about how PowerPoint can have the effect of "dumbing down" debates have been developed by the prominent academic Edward Tufte. See, for example, his whimsically titled but quite serious article "PowerPoint Is Evil" in the September 2003 issue of *Wired* magazine.

DVIC DD-SD-07-15617

Gen Tommy R. Franks, Commander in Chief, Central Command, and Donald R. Rumsfeld, Secretary of Defense, brief reporters at the Pentagon. Gen Franks and Secretary Rumsfeld worked closely in planning the ground campaign in Iraq.

not need to send 500,000 troops across the line of departure to do the job.[87]

Secretary Rumsfeld and General Tommy Franks worked closely on the development of the plan, which stemmed in part from the fact that the combatant commanders reported to the Secretary of Defense (and certainly not to any of the Service chiefs, as General Franks made very clear in his memoirs, with its colorful, "aw, shucks, I am just a country boy" language.) It was also a function of Rumsfeld's hands-on style and his determination to implement his vision. In his memoirs, General Franks described frequent personal meetings, and almost daily telephone contact, between the two men. Franks would propose a course of action. The secretary would react to the proposal and ask him to come up with a new course of action in a few days, at most a week or two, or to provide additional details. This set a grueling pace. Although General Franks concluded that the final product was all the more robust on account of the process, the short-term effect of the secretary's input was to place additional burdens on the planning staffs at CentCom and its subordinate commands like MarCent and I MEF.[88]

Even without PowerPoint or the secretary's input, planning the ground war would have not have been easy. Joint planning on a large scale is just plain hard, and it is not something that happens often, especially for Marines. Most Marine deployments and exercises (with the notable exception of Ulchi Focus Lens in Korea) were on a much smaller scale, which was fine with the Marine leaders who believed that the Corps should focus its efforts on preparing for real-world, Marine expeditionary brigade-sized commitments while the expeditionary force performed the Title 10, U.S. Code functions (of organizing, equipping, and training the force).[89] Not that it was much easier for Army officers. They were certainly more used to operating on a large scale—the Army was still preparing its officers to staff and fight at the corps level, as it had since before World War II—but as General McKiernan pointed out, this was the first time in decades that a CFLCC would plan and fight a combined/joint operation above the corps level.

The starting point for understanding the plan for the ground war was Desert Storm in early 1991, when General H. Norman Schwarzkopf had painstakingly assembled a Coalition of over 500,000 troops in and around Saudi Arabia to expel the Iraq invaders from Kuwait. No one moved north until all of the troops were in theater, equipped and ready to attack. After the air campaign that lasted over a month, more than 15 United States and allied divisions advanced into Kuwait and Iraq. The ground war lasted only some 100 hours before the United States announced a ceasefire, Kuwait having been liberated. With the exception of the Kurds in northern Iraq, who came under the protection of a combined/joint task force in Operation Provide Comfort in the spring of 1991, the Iraqis were more or less left to their own devices. This had devastating consequences for the Shi'ite rebels, especially those in the South, who had dared to rise up against Saddam in the confusing days that followed the ground war.[90] It was Saddam's suppression of the Kurds and the Shia that had led to Operations Southern Watch and, later, Northern Watch, to enforce the no-fly zones in the southern and northern thirds of Iraq through 2003 and keep the Iraqi military in check.

The planners of Desert Storm would have been comfortable with the CentCom plan that was on the shelf for war with Iraq, Operations Plan (OPLAN) 1003-98. The premise of the plan was that Iraq had once again attacked Kuwait, and that CentCom had come to its defense and counterattacked, its objective being the removal of Saddam Hussein and his regime. It was a relatively heavy plan, with five divisions crossing the line of departure with I MEF supplying the command element.[91] In late 2001, before focusing on Afghanistan, CFLCC planners had considered a limited objective plan, an attack to seize the oil fields in southern Iraq, which did not take as its starting point OPLAN 1003. After they were tasked in early 2002 to come up with a more ambitious plan, they still did not use 1003 as the starting point but developed new courses of action. The course of action favored by General Mikolashek, the CFLCC com-

Photo courtesy of VMFA(AW)-121

A Lockheed KC-130 from Aerial Refueler Transport Squadron 352 refuels two McDonnell Douglas F/A-18s from Marine All-Weather Fighter Attack Squadron 121 as they soar high over Kuwait during one of the many 3d Marine Aircraft Wing deployments to the Middle East in support of Operation Southern Watch.

mander at the time, was in line with CentCom's concept of "shock and awe," emphasizing speed and surprise. It was an audacious plan calling for the force to avoid the lengthy build-up that would announce United States intentions and, using prepositioned equipment, launch one division deep into Iraq with a view to reaching Baghdad within 10 days.[92]

Coalition Forces Land Component Command took this plan on the road and through all of the various personal and impersonal contacts, including the planning conferences of various sorts, it evolved into something not more but less audacious that came to be known by the rather nonmilitary phrase "generated start." This was a throwback to Desert Storm and OPLAN 1003 in that it called for a deliberate build-up of decisive strength before anyone crossed the line of departure. The build-up would take approximately ninety days, and the simultaneous attack by two corps would take up to ninety more days to get to Baghdad.[93] The results of the first I MEF operational planning team, which ran from 13-19 March at Camp Pendleton, further illustrate both the plan's general outlines and I MEF's role in it. The planning team assumed that I MEF would be a supporting effort, with the Army's V Corps (usually based in Germany) as the main effort. I MEF would deploy 1st Marine Division, 3d Marine Aircraft Wing, 1st Force Service Support Group, and a regimental combat team out of II Marine Expeditionary Force (II MEF) to serve as "Task Force South." The team came up with a concept that called for the division to attack northwest from Kuwait to seize the airfield at Jalibah, which would become a support area, including a forward arming and refueling point, before proceeding to seize the airfields in the vicinity of Qalat Sikar and Al Kut to the northeast, much closer to Baghdad. At this point they expected CFLCC to order a significant pause to allow the Army to flow additional forces and supplies before continuing the attack toward Baghdad. The pause was an idea that never sat well with the Marines. In the meantime, Task Force South would resume responsibility for southern Iraq, addressing any threats the division had bypassed and in general securing support areas and lines of communication.[94]

This did not play well in Washington. Considering

the possible scenarios that could lead to war with Iraq, few at the Pentagon were willing to accept a plan that took 90 days to launch. The word came down that Secretary of Defense Rumsfeld was unhappy with the ponderous nature of the "generated start," which would not only be logistics intensive but would also sacrifice any hopes of surprising the enemy. The result was that CentCom felt pressure to develop ever lighter, more creative scenarios.[95]

The warfighters moved the plan in the direction that Rumsfeld indicated. By early summer, the pendulum had swung so far to the "light" that there was a plan under consideration for something like two brigades, perhaps one Army and one Marine, to cross the line of departure soon after a *casus belli* had occurred. To be sure, this would be a "running start"; this small force would be reinforced by heavier forces from the Army and Marine Corps as soon as they could get to the fight. The running start seemed to mean starting early but finishing late, since the force would once again have to wait for reinforcements before moving north for the decisive battle of Baghdad. Marine planners pointed out that there was no need for two corps headquarters to control the running start, and argued that I MEF could provide command and control for the initial phase of the operation. It would be better suited for the task, because, unlike V Corps, it not only had organic subordinate commands, but it was also experienced in running and controlling air operations, which V Corps was not. At least initially, the Army was reluctant to accept this argument.[96]

The pendulum started to arc back toward the middle in the second half of 2002, toward the ultimate, or "hybrid," version of the plan, a mix of elements from the "generated start" and the "running start." One of the problems with the running start turned out to be that it was very difficult to come up with an optimal time to attack, considering the trade-offs between strength and surprise. When would the mix be right, early enough for surprise but "late" enough to allow CFLCC to build up the strength that it needed to do the job? Over the summer, some of the planners considered deploying a stronger initial force, perhaps two Army brigades and two Marine regimental combat teams that could attack 30 days after the force flow started. An attractive feature of the stronger force was that with it CFLCC could attack simultaneously in two directions, with the Army attacking to the northwest, toward the city of An Nasiriyah, and

East Coast Marines in a West Coast Plan

What became Task Force Tarawa started as Task Force South. The concept for a task force to follow in trace of a division, to be charged with neutralizing any threats that division had bypassed and then securing I MEF's lines of communication, was certainly on the table by March 2002. At a time when the Pentagon wanted to keep the force light and Headquarters Marine Corps was thinking in terms of "global sourcing," that is, pulling assets from all over the world to meet the potential need for Marines in CentCom, it made sense for "the Commandant's G-3" at headquarters to look to II MEF in Camp Lejeune, North Carolina, to supply something like a brigade-sized force to float to theater on the amphibious task force that was based on the East Coast. The location and capacity of amphibious shipping helped to determine where the forces would come from and how big they would be. Since there was another amphibious task force on the West Coast, the Marine Corps could use the task forces to get two separate elements to the fight quickly and securely. The East Coast Marines could mount out of Morehead City, North Carolina, on to their own amphibious task force without adding to the long list of units for deployment from the West Coast, and the two forces could travel by different routes. Another reason to look to Camp Lejeune was that the only other remaining brigade-sized force was in Okinawa. That force needed to stay there in case it was needed in Korea. At this point the planners were looking at a brigade-sized unit that would not fight as a Marine air-ground task force. It might deploy without the expeditionary brigade command element. If that happened, all of its parts could be parceled out after arrival in theater. Whether to include a command element with the East Coast contingent was a matter mostly for the I MEF commander to decide. This would happen only after General Conway took the helm in the fall of 2002.*

*LtCol George W. Smith, Jr., intvw, 8Jun04 (MCHC, Quantico, VA); Col Nicholas E. Reynolds, "OIF Field History Journal," 2003, entry for 1Jul04, reporting the comments of Col Ronald J. Johnson, Headquarters Marine Corps (PPO) current operations officer and Task Force Tarawa operations officer.

the Marines attacking to the northeast, toward Basrah, Iraq's second city, and the nearby southern oil fields.[97] The Marines had mixed feelings about Basrah, especially the suggestion that they seize and run Basrah International Airport, no doubt an important symbolic target but one that could become a "force sump," because to run the airport they would have to protect it from portable antiaircraft rockets, and that would mean controlling the city, which could take a regimental combat team a long time.[98] By late July, the Army had accepted the proposition that I MEF would control the initial phase of the war, roughly the first 75 days, and that the 3d Infantry Division would be under I MEF control at least for that period. There was further discussion of a lengthy (30-45 days) pause in the middle of the campaign to build up supplies for the final assault, a proposition that had not become any more attractive to the Marines. Finally, there was discussion about another large city, Baghdad, but there would not be anything like a final decision about how to tackle the capital until much later.[99]*

September saw a declaration by General Franks that President Bush wanted CentCom to be prepared for war within 60 to 90 days and that CFLCC needed to be prepared to execute across the continuum of force, with either a heavy or a light plan. During the same month, I MEF began to work with British Royal Marine planners on their potential role in an offensive; initial plans called for one Royal Marine commando (the rough equivalent of a Marine Corps battalion) to fight under I MEF. The Royal Marines appeared ideally suited for an amphibious assault against targets on the Al Faw Peninsula in southeastern Iraq.[100]

The American Marines took an immediate liking to their British counterparts, as individuals and as warriors, and welcomed them as reinforcements. This was eventually to lead Colonel Christopher J. Gunther, who became the I MEF plans officer in the second half of 2002, to take advantage of a brief to the British high command to ask for a second Royal Marine commando, tossing in as a sweetener the possibility that a U.S. Marine expeditionary unit might work under British command.[101]

Most versions of the plan included a healthy dose of "deep" and special operations against carefully selected targets, like those seen in Afghanistan, which had impressed General Franks favorably. There would be two joint special operations task forces,

*Both I MEF and V Corps planned extensively for the fight even after CFLCC decided in September that Baghdad planning would fall to V Corps.

JCCC 030127-M-2081S-001

Newly arrived Marines prepare a bivouac area near the command post center at Camp Commando, Kuwait.

Joint Special Operations Task Force North and Joint Special Operations Task Force South, the first to work with the anti-Saddam militias in Iraqi Kurdistan in the northeast part of the country, and the second to secure any potential launch sites for Scud missiles in the western desert between Baghdad and the Jordanian border (which had been used by the Iraqis in Desert Storm to attack Israel). But that was not to be the extent of special operations; there would be special operations force elements at work throughout the country, adding their capabilities to the force mix in all areas of operation. Marine expeditionary force planners took the initiative to coordinate with special operations forces, for example, by traveling to Fort Campbell, Kentucky, for an unexpectedly cordial and productive session. Once again, the Marines found that they continued to get good results when they reached out to other communities.[102]

In October, I MEF held a general officer symposium to develop "a common understanding of the . . . operation plan." It would have been more accurate to say the state of "operational planning," since "the" plan did not yet exist–even though it was now beginning to be known as Operations Plan 1003V. During the same month, there was a I MEF exercise based on an invasion of Iraq, and advance parties from I MEF began to deploy to Kuwait to stand up a command post, soon followed by the I MEF command element, which deployed to Camp Commando, a few miles west and north of Kuwait City, and prepared to participate in CFLCC exercises. By late November there were approximately 850 members of the force staff in Kuwait, along with approximately 100 members of the 1st Marine Division staff.[103]

As the commanders and their staffs began to settle in, or at least try to settle in, during December, they realized they could be living in the desert for quite some time. Despite the pronouncements from CentCom, there was still nothing like a timeline, no one knew for sure when the war would start, or even if it would start. On 9 December, General Conway spoke to his officers at Camp Commando and told them they might be in Kuwait for the long haul. He wanted to dispel rumors that the Marines were going home after the current round of exercises. He did not know if there would be a war but ventured the guess that it might come "as late as February," which was still during the "cool" season, the "right" time to go to war. But he concluded that if the Marines had to fight in the heat, they would fight in the heat.[104]

Around the same time, senior Marine officers debated whether they should push for a large-scale deployment. The issue was this: if I MEF arrived early, in force, a large number of Marines might spend a number of months in Kuwait waiting for a war that might never come. On the other hand, if the troops stayed home until the situation gelled, they might not be able to get to theater in time if the pace of events were to speed up and the war started on short notice. General Mattis, supported by Colonel Gunther, is said to have pleaded the case for flowing the forces in the near future. Colonel Gunther pointed out that at worst the Marines would get some good training; it would be like going to a very big series of combined arms exercises in a desert other than the Mojave at Twentynine Palms, California.[105]

GySgt Jay R. Joder, an intelligence specialist, works in the Coalition Forces Land Component Command's operations and intelligence center at Camp Doha, Kuwait. Ground operations within Central Command's area of responsibility would be controlled from here.

Photo courtesy of CFLCC

In the meantime, General McKiernan, who had replaced General Mikolashek in September, was working to turn CFLCC into a more joint command. McKiernan liked to say that CFLCC was made up of a number of "tribes" and "sub tribes," that is members of different Services, American and allied, as well as members of different branches of the Army. Many Marines forget the extent to which Army officers often identify with their military occupational specialty. His method was to do "a lot of training and tribal team building." One McKiernan initiative was to reorganize the staff along functional lines in order to break down the old-fashioned "G-1, G-2, G-3, G-4" stovepipes and encourage cross talk among the various disciplines. Virtually every day in late 2002 and early 2003 the CFLCC staff held some sort of exercise, in addition to participating in CentCom exercises and various rehearsals. This may turn out to have been one of the most extensively rehearsed wars in U.S. history. In short, the point was to change the original tribal identity, to make CFLCC into a joint tribe, a headquarters with a joint outlook.[106]

General McKiernan did not just want to change the tribe's way of thinking, he also wanted to change its makeup. In the fall of 2002 there was also a push to get the other Services, especially the Marine Corps, to send more officers to CFLCC. The long-standing joint manning document called for a complement of some 90 Marines; in the early fall there was only a handful of Marines at Camp Doha, ordered to CFLCC by MarCent. One or two had even been at Doha since the fall of 2001 and were filling key billets. Two of these officers were Colonel Gregory J. Plush whose title was MarCent liaison but who functioned for months more or less as the senior Marine on the staff, and Colonel Marc A. Workman, chief of the deep operations coordination cell in the C-3. General McKiernan relied heavily on Workman to plan what was known as the "deep fight," operational fire support well beyond the frontlines, which at CFLCC in 2002 and 2003 was about "creating desired effects," not just destroying targets.[107]

What Marines do for Christmas: 25 December at Camp Commando

Commando sat at the base of Mutla Ridge, one of the few parcels of distinguishable terrain in an otherwise flat and largely treeless country. In August 1990, the Iraqis had used the ridge to shell the camp below; there were still shell holes and shattered buildings here and there within its confines. For a Christmas treat, I MEF Headquarters Group gave its members a rare opportunity to venture outside Camp Commando on a 10-kilometer run to the ridge and back and to find out that they were not in the same shape that they had been in at Camp Pendleton. Armed guards were posted along the route to defend against possible terrorist attacks. Even with the terrorist threat, and the sand that offered little traction, it was an enjoyable outing. Back at Camp Commando, Generals Amos and Mattis spent their time talking about Baghdad with a handful of I MEF officers. General Mattis said he wanted to get north of the Tigris as quickly as possible to minimize the time his division would be susceptible to weapons of mass destruction; General Amos said he wanted to put the wing on a "JDAM [or "smart" bomb] diet" so that he would have enough precision munitions for the big fight in the city. Then they ranged over a variety of issues–from where to place boundaries between Army and Marines; how to coordinate fires and control air space; the integration of information and covert operations on the one hand and conventional operations on the other; and general techniques, tactics, and procedures for fighting in urban terrain. One of the participants, Lieutenant Colonel George Smith, remembered thinking that this "pure warfighting" talk was the ultimate professional military education session.*

*Col Nicholas E. Reynolds, "OIF Field History Journal," 2003, entry for 19Jul04; LtCol George W. Smith, Jr., intvw, 8Jun04 (MCHC, Quantico, VA).

Photo courtesy of Field History Branch

Photo courtesy of CFLCC

MajGen Robert R. Blackman, Jr., chief of staff for the Coalition Forces Land Component Command, talks with the command's logistical officer following one of many updates at the Coalition Operations and Intelligence Center, Camp Doha, Kuwait.

McKiernan was so pleased with Workman, in particular his experience as a Marine air-ground task force planner able to coordinate a range off effects, that he resisted any moves to replace Workman, who wound up serving at CFLCC for the duration, working up to 18 hours a day for months on end. It was only after they had been working together for some time that Workman mentioned he was a rancher by profession, and not a regular officer, which McKiernan at first refused to believe, a backhanded compliment to the professionalism of this Reserve artilleryman who literally had to sell the farm in order to stay on active duty.[108] In October, reinforcements for Plush and Workman began to trickle in. General McKiernan asked for the best Marine general that Headquarters Marine Corps could find, and Major General Robert L. Blackman soon received orders to serve as the chief of staff at CFLCC. General Blackman, who had commanded the 2d Marine Division and most recently served at CentCom headquarters in Tampa, was respected both as a Marine and as a team player in the joint arena.[109] After his arrival in country, General Blackman repeatedly urged Headquarters Marine Corps to provide more officers for the staff. In late November, Colonel Patrick J. Burger arrived at Doha to serve as the senior I MEF liaison officer and began to coordinate the activities of the growing number of Marine liaison officers who were not technically part of the CFLCC staff but who in most cases might as well have been.[110] In January, during a trip to Washington, D.C., General McKiernan himself again requested Marine augmentation for the CFLCC staff. It is hard to escape the conclusion that if CFLCC was short of Marines, it was a reflection of Marine manpower constraints and certainly not of a desire by "Big Army" to limit the number of Marines at CFLCC or the influence of Marine doctrine.[111] In the end there were some 70 Marines at CFLCC, still short of the number in the joint manning document that the Army and the Marine Corps had long since agreed to, but certainly at a level where the liaison officers could effectively represent the Marine point of view. More importantly, the staff became even more joint in the interests of putting the common interest ahead of any one Service or nationality. In February 2003 there would be some 1,300 members of the CFFLC staff. [112]

By November both the CFLCC and the I MEF staffs

Photo courtesy of Field History Branch

Marine Lieutenant Colonels Robert L. Sartor, current files officer, and Brian D. Kerl, assistant current files officer, both with the operations section, I Marine Expeditionary Force, exchange information during Exercise Internal Look 03. The bi-annual exercise was designed to evaluate the command and control capabilities of Central Command headquarters and its component commands.

were ready to exercise the most recent version of the CFLCC portion of the CentCom plan, which now bore the name "Cobra II" and still called for a relatively small force to advance from Kuwait into southern Iraq. Major Evan A. Huelfer, USA, FLCC's lead planner, commented that although there were some changes, the outlines of the "four brigade" initial attack had not changed markedly over the intervening months.[113]*

What was somewhat different in the fall was the hope that the U.S. Army's 4th Infantry Division, augmented by a British division, would advance from Turkey into northern Iraq, toward Baghdad, at roughly the same time.[114] Generals Franks, Mikolashek, and McKiernan all wanted near-simultaneous attacks into Iraq from north and south, the main attack coming from Kuwait, with a supporting attack by Coalition forces from Turkey. This was an idea that certainly sat well with the outgoing commandant of the Marine Corps and incoming head of European Command, General James L. Jones, Jr., who wanted to do what he could to make it happen, and was willing to commit Marine units if it made sense to do so.[115] The attack from the north would be able to secure important targets in the north such as the oil fields around Kirkuk and Saddam's "hometown," Tikrit, which was considered to be one of his bases of power.*

"Lucky Warrior 03-1" was a CFLCC exercise testing command and control, including the links between CFLCC and I MEF, and focusing on the initial phases of war. It began on 24 November, just in time for Thanksgiving, and was a precursor to the CentCom exercise "Internal Look 03," which tested three scenarios from Operations Plan 1003V at the next higher level and occurred in early December with I MEF participation. General Blackman remembered later that the scenario at the time was for CFLCC elements to cross the line of departure with two brigade com-

*Both Gen McKiernan and his special assistant, Terry Moran, remembered a version of the plan in the fall of 2003 around the time of their arrival at Camp Doha. The general termed it "a very small force . . . a brigade combat team from the 3d ID and a MEU from the Marines." Moran used more or less the same terms: "Early on, there was some discussion of that start force being no more than a brigade of the 3d ID, reinforced out of the MEU . . . so the thought was that we would cross the LD with that very small force, and it would be rapidly reinforced by the MEF and by the V Corps. That was probably a bit . . . imprudent." (LtGen David D. McKiernan intvw, 30Jun03 [U.S. Army Center of Military History, Washington, DC]; TerryMoran intvw, 23Aug03 [U.S. Army Center of Military History, Washington, DC])

*The history of this idea goes back to the spring of 2002, when Gen Mikolashek proposed it to Gen Franks, who accepted it. However, it was not actively pursued until the summer of 2002, when it was briefed to the British, who were very enthusiastic about the idea and played a key role in resurrecting it. British sources are consistent with Huelfer's memory. (Maj Evan A. Huelfer intvw, 16Mar03 (U.S. Armt Center of Military History, Washington, DC); House of Commons Derfense Committee, *Lessons of Iraq* [London, UK: Stationary Office, 2004] v. 1, p. 45.)

bat teams from 3d Infantry Division and one Marine expeditionary unit, all under I MEF. The expeditionary unit would take the southern city of Umm Qasr, while the two brigade combat teams would take the oil fields. There would of course be follow-on forces under this "running start" scenario.[116*]

After these exercises, General McKiernan decided he needed a stronger force that could move fast; unless CFLCC were stronger from the outset, he would not be able to accomplish all of the set tasks at the same time. It seemed, for example, that CFLCC would have to choose between maintaining security on the southern oil fields and moving north. As McKiernan put it on 19 December 2002: "We do not have enough combat power to simultaneously penetrate [Iraq] and move straightaway . . . to Baghdad and do all the other tasks we have to do in southern Iraq . . . secure the oil fields, keep Basrah out of the fight, develop our logistic support areas, deal with displaced civilians, deal with enemy prisoners of war, cross the Euphrates."[117]

After the war, on 1 May 2003 General McKiernan made the comment that this was "probably the most critical decision I made [during Operation Iraqi Freedom] . . . I made a case for additional forces . . . that were ready to conduct ground operations . . . [for a] two-corps operation, with a penetration as a form of maneuver by the main effort, and a [simultaneous] supporting attack by the MEF to fix everything over on the east side."[118]

Next there was a series of video teleconferences between General McKiernan and CentCom so that General McKiernan could make his case. In Major Huelfer's words, it went something like this:

> Lieutenant General McKiernan made a plea to General Franks, and said, "Hey, look, you are asking to attack at C+15 [15 days after starting the force flow], but we won't have enough forces on the ground to do both of those operations simultaneously. We cannot go out to Nasiriyah and toward Basrah at the same time." . . . So . . . I think the light went on, he [General Franks] said, "Okay.". . . The very next day, General Hailston [the MarCent commander] came up . . . to Kuwait, and the next thing you know . . . [Marine] regiments were flowing into Kuwait.[119]

In fact, there was an intermediate step. The talks between McKiernan and Franks led to talks between theater and the Pentagon, a sometimes painful process described by McKiernan's assistant, Terry Moran (no doubt with tongue in cheek), as a "tug-of-war": "We had a series of VTCs [video teleconferences] where we were asking to move the force posture from X to X plus Y. We were in a tug-of-war with CentCom and with OSD [Office of the Secretary of Defense] on how big Y could be." It seemed that whatever increase CFLCC requested, "they would try to skinny [it] . . . down."[120] Nevertheless, the end result was that CFLCC came to develop the heavier, two-corps plan, which the 3d Infantry Division reverted to V Corps.[121]

This plan was truly a hybrid, a good combination of the "generated start" and the "running start." The force was by far smaller, and more agile, than the force with which General Schwarzkopf crossed the line of departure in 1991, but it was developing a respectable amount of combat power. Where Schwarzkopf had had two Marine divisions at his disposal, it was now fairly official that if and when the order was given, I MEF would bring its entire division-wing-force service support group team along with some 7,100 Marines from Camp Lejeune. General Conway decided he wanted them to come with the 2d Marine Expeditionary Brigade command element; he wanted a second Marine command element in country to take on the distinct security and stability missions in southern Iraq that had originally been foreseen for Task Force South a few months earlier.[122*] I Marine Expeditionary Force's subordinate el-

*The exercise was "McKiernan's first opportunity to . . . conduct operations with his new staff and new general officers and to exercise the new [CFLCC] organization . . . [as well as] the first opportunity for CFLCC's major subordinate elements . . . to practice operations under . . . CFLCC. Much of the exercise focused on team building and establishing [joint] standard operating procedures." (Gregory Fontenot, et al., *On Point: The U.S. Army in Operation Iraqi Freedom* [Fort Leeavenworth, KS: Combat Studies Institute Press, 2004], p. 53)

*Part of the explanation for the selection of 2d Marine Expeditionary Brigade as opposed to 2d Marine Division may have been that much of the division staff was already committed to an ongoing CentCom operation in the Horn of Africa and was simply not available. The 7,100 Marines would include the three infantry battalions of Regimental Combat Team 2 (RCT 2), something less than 1,000 Marines in the brigade headquarters, a combat service support element, and 81 aircraft. Apart from the units that were organic to 2d Marines, the larger units earmarked for deployment with the brigade were 2d Battalion, 8th Marines, which replaced the already deployed 2d Battalion, 2d Marines; 1st Battalion, 10th Marines; 2d Force Reconnaissance Company; Company A, 8th Tanks; Company A, 2d Assault Amphibious Battalion; and Company C, 2d Light Armored Reconnaissance Battalion. Upon arrival in Kuwait, the aviation and support elements would transfer to the wing and the force service support group, but later on other units came under Tarawa's control, such as 15th Marine Expeditionary Unit, enabling it to reach a maximum strength of some 15,000 Marines and sailors.

How Task Force Tarawa Got Its Name

General Conway did not like the colorless designation "Task Force South" for 2d Marine Expeditionary Brigade, and he asked Brigadier General Richard F. Natonski, its commander, to come up with something better. A Marine with a good sense of history, General Natonski wanted to think the matter through and began by asking the History and Museums Division for help in naming the task force. The head of Reference section, Danny J. Crawford, reviewed the names of the task forces in Desert Storm, such as Grizzly, Papa Bear, and Ripper, and made a few suggestions for the new task force—Mameluke, from the name of the sword; Fortitude, one of the Corps' earliest mottoes; Chosin or Chosen Few, to commemorate the 50th Anniversary of the Korean War (which had ended in 1953). Crawford added the incidental note that the Navy's codename for Iraq in World War II had been "Plughole," clearly a non-starter but an interesting bit of historical trivia. General Natonski agreed and said he would consider Crawford's other suggestions along with the idea that he pick the name of a mythical creature from the Arab world that his future opponents would recognize. In the end, Tarawa seemed to fit best. General Natonski happened to be looking at some artifacts from World War II in the Pacific, a reminder that Tarawa was a famous battle associated with 2d Marines and therefore a good name for a task force with that regiment as its infantry. When 2d Marine Expeditionary Brigade landed in Kuwait, it became known by that illustrious place name.*

*BGen Richard F. Natonski intvw, 26Mar04 (MCHC, Quantico, VA); Natonski-Crawford e-mail msg, 18Dec02 (RefSec, MCHC, Quantico, VA)

ements, then, were to be 1st Marine Division, 3d Marine Aircraft Wing, 1st Force Service Support Group, Task Force Tarawa, and I MEF Engineer Group, a force-level engineering asset. The last was a concept that was unfamiliar to many Marines. Made up of Marine engineers and Navy Seabees under the command of a rear admiral, the engineer group was an initiative that dated back to the days when General Zinni was the force commander. The point was for I MEF to be able to do "one-stop shopping" for its engineering needs, especially in major deployments.[123]

In December 2002 and January 2003, the rough outline of the "hybrid" plan was for the Army's V Corps to move north through the western desert of Iraq with at least one full division, with the ultimate goal of capturing Baghdad. Meanwhile, I MEF would move north on a more central axis, the "obvious" route to Baghdad, which leads into and through the heart of Iraq, the Fertile Crescent between the Tigris and Euphrates Rivers. The Army remained the main effort, the Marines were the supporting effort, which meant they would pick as many fights as they could, as General Conway put it, in order to deflect pressure from the Army and to make the Iraqis think the Marines were the main effort. In effect, the Marines would defend the Army's flanks by rapidly defeating enemy forces in its zone. I Marine Expeditionary Forces' spearhead division, would have the city of Al Kut as its aiming point, while Task Force Tarawa would seize and hold objectives in the south, securing the division's rear from potential threats.[124] A quick look at the map reveals that Task Force Tarawa was to assume responsibility for at least one major city, Basrah, the surrounding oil fields, and thousands of square miles of territory, much of it inhabited—a breathtakingly ambitious mission that could succeed only if the optimistic assumptions about the nature of the opposition (or the lack thereof) were correct.

Exactly what the Marine Corps would do in Baghdad was once again unclear. Along with the elite forces that would most likely try to keep the Coalition away from the capital, Baghdad had long been recognized as the enemy's center of gravity. The Marines wanted to play a role in the fight for Baghdad and argued for a simple boundary, like the Tigris River, between Marine and Army units in the zone. By late summer 2002, the CFLCC commander, General Mikolashek, appeared ready to accept that approach. Then, in September his successor, General McKiernan, decided to change course and designated the V Corps staff as the lead on Baghdad. The Army was the main effort and it had the heavier force. It would fight in the city; the Marines would support the attack but remain outside the city limits. To be sure, there was always the possibility that this would change, both in General McKiernan's mind and in the minds of the Marine planners. To the general's way of thinking, the "base plan" was for V Corps to command the tactical fight for Baghdad. But "there was a branch plan that said, if . . . it makes sense . . . bring forces

DVIC DM-SD-05-04662

A graduate of the University of Louisville with a degree in history, BGen Richard F. Natonski, following a tour with Plans, Policies, and Operations Division of Headquarters Marine Corps, assumed command of 2d Marine Expeditionary Brigade (Task Force Tarawa) in June 2002.

into Baghdad from both the . . . MEF . . . [and] . . . V Corps, . . . we ought to be sure that we can do that."[125]* The bottom line was that both V Corps and I MEF continued to think about Baghdad and to try to come up with the right way to subdue and control the Iraqi capital.

The prospect of a joint British-American attack from the north lasted through most of December 2002, when it became increasingly clear that the Turks were not likely to allow the British to pass through their country. It was only later that the Turks also refused to allow an American division to pass through Turkey.** In January 2003 it was easy for CentCom and CFLCC to decide to redirect as many British forces as possible to Kuwait and to I MEF, which did what it could to encourage this development. The reinforcements came in the form of the 1 United Kingdom Division (Armored), a composite made up of three more or less independent brigades, which began to flow into Kuwait in early 2003. Marine expeditionary force planners happily proceeded to include the British division in their plans, along with the 15th MEU (SOC), which was now officially slated to reinforce the British.[126]

With a full division at their disposal, the British could assume responsibility for Basrah and its surroundings. In the words of one staff officer, Lieutenant Colonel Richard T. Johnson, who worked on the plan at CFLCC, the Marines "hit the right numbers" (like a lucky gambler) when the British joined their team. Johnson's thinking, in retrospect, was that without the British, the Marines would have been hard pressed to control the powder keg that was Basrah and accomplish their other assigned missions.*

*This is one of the few instances when there was a whiff of Service politics in Gen McKiernan's basic policy decisions. There was no compelling reason to assign Baghdad exclusively to V Corps.

**It would not become clear until mid-March 2003 that the 4th was not welcome in Turkey either. By then the U.S. Army's 4th Infantry Division and its equipment were in ships floating off the coast of Turkey. But this was not all bad. It enabled Gen Franks to bolster a deception plan that pointed to an attack from the north, which may have found its mark. (Tommy Franks, *American Soldier* [New York, NY: Regan Books, 2004], p. 429)

*LtCol Richard T. Johnson noted that in December 2002 the senior British advisor at CFLCC, BGen Albert Whitley, met with Gen James Conway to discuss whether and how the British Army and I MEF could work together. (LtCol Richard T. Johnson intvw, 26Apr03 [MCHC, Quantico, VA])

JCCC 021112-N-4374S-020

Marines assigned to the 24th Marine Expeditionary Unit (Special Operations Capable) disembark from an amphibious assault vehicle to conduct a live fire training exercise while on their six-month training deployment in the Central Command area of responsibility.

Another planner, Lieutenant Colonel Smith, said the British were I MEF's "slingshot," propelling it much farther and faster than it could otherwise have gone.[127] With the British in I MEF, Task Force Tarawa's mission changed; its focus could shift westward, away from Basrah and the southern oil fields. But the type of mission remained similar—it was to secure crossing sites over the waterways in the south and to preserve combat power. General Natonski stated that he became aware of this change on approximately 17 January 2003.[128]

There were similarities, and dramatic differences, between the battle space assigned to the Marines and that assigned to the Army. Most of the Army's battle space was trackless desert, at least until it reached the cities of Najaf, Karbala, and Hillah, which lay southeast of Baghdad. Being heavily mechanized, the Army was, arguably, well suited for the desert. The Marine battle space, on the other hand, started off as desert but soon became a varied and complex mix of desert, scrub brush, agricultural land, rivers, and canals. It was defined, and largely contained, by the two great rivers, the Euphrates and the Tigris, and anchored in the south by Basrah, which lay on the Shatt al Arab waterway not far from the border with Iran, and in the north by Baghdad, some 300 miles to the northwest. A small number of highways ran through the area, varying in quality from a good superhighway to two-lane local highways to dirt roads. Like the rivers, the highways mostly ran on a northwest to southeast axis. There was a prominent chokepoint at An Nasiriyah in the south of Iraq, some 90 miles west-northwest of Basrah, where three highways, one river, the Euphrates, and a canal came together. It was on the way for anyone going north or west from Kuwait or Basrah. The argument has been made that the Marine battle space was a good fit for the Marine Corps' capabilities, being closer to the ocean, calling for the ability to cross bodies of water, and encompassing cities that a Marine regiment, with its many infantrymen, could subdue more readily than an Army brigade, which had relatively few "dismounts."[129]

The I Marine Expeditionary Force took CFLCC's "Cobra II" Operations Plan and used it to develop its own plan, which had four phases and came to include various branches and one sequel. Similar to contingencies that could become part of the plan, the branches included were early military collapse, a variant of "catastrophic success," inundation, which could occur if Iraqi engineers intentionally flooded parts of the region between the Tigris and the Euphrates, and the seizure of the crossings over the Euphrates in and around An Nasiriyah. One sequel, a mission that I MEF could receive after completing the missions in the base plan, addressed the seizure of the oil fields in the north around Kirkuk. The Marines did not view all of the branches and the sequel as equally likely, or any one of them as particularly likely, but wanted to err on the side of being prepared. Speaking about branch plans at his level, General McKiernan referred to "the old adage that you can have a great plan, but the plan [c]ould change at the line of departure for a variety of [reasons] . . . that we don't have a great deal of control over. . . . [W]e ought to . . . have lots of options. . . . [W]e can decide which one we are going to execute to obtain the initiative throughout the fight. . . . [W]e need to go through the planning of different branches . . . so they don't become surprises to us." The general stressed that it was important for the various parts of the joint force to coordinate their branch plans in advance.[130]

The four phases of the plan ranged from reception, staging, onward movement, and integration, commonly abbreviated RSOI, a needlessly complicated acronym for Phase I, or "preparation," which was mostly about deploying the forces to theater and preparing them for combat, to "shaping the battle space" through preliminary attacks in order to degrade Iraqi command and control, and seize key pieces of terrain. This was the phase when various special operations troops under CentCom's Joint Special Operations Command would engage the enemy, especially in the western and northern parts of Iraq, and there would be air attacks in the northern and southern no-fly zones, which had already done much to ensure air supremacy for the Coalition. Then there was Phase III, which included the main air offensive and decisive ground maneuver. Finally there was Phase IV, post combat operations, which encompassed security and stability operations.[131]

Phase III was the heart of the plan; there was amazing breadth and depth to the parts of the I MEF plan dealing with this phase. Phase IV, in comparison, received very little attention.* There was very little guidance from higher headquarters on Phase IV, not even a basic policy decree. Some of the I MEF planners found this troubling and got out ahead of their higher headquarters, making preliminary plans for Phase IV on their own. They realized that, like it or not, the Marine Corps would be involved in Phase IV operations, though hopefully for only a relatively brief period of time. This was, after all, the kind of operation that had traditionally been left to the Army. But as General Conway put it later on, the Marine Corps had "always done windows" and would now do whatever the President, Secretary of Defense, or combatant commander directed.[132]

A maneuverist, especially if British, might argue that there was too much detail in Phase III of the I MEF plan; it was reminiscent of the bad old days when the Marine Corps prepared "to fight the plan" as opposed to the enemy. In this conflict, British plans tended to be very brief and to the point in comparison with American plans. One British planner, Lieutenant Colonel James Hutton, Royal Marines, could not believe all of the time he spent at planning meetings with his American counterparts, especially the U.S. Navy Sea-Air-Land (SEALs) personnel, who, he thought, tried to plan for every last possibility. It was, he said, "mind-numbing" and inhibited flexibility. He added the thought that the Royal Marines might appear slack by comparison, but they also felt they had more leeway to react to situations as they developed.[133] When asked about this topic, the typical I MEF staff officer occupied the middle ground between the SEALs and the Royal Marines, believing that the plan itself was nothing but that planning was everything, because it forced the operators to prepare for a broad range of contingencies. As one of the lead planners, Colonel George F. Milburn, I MEF's Future Operations officer, commented: "[Y]ou have

*As one senior Marine operations officer put it very forcefully before the war, there was "absolutely no plan for Phase IV. None. Zero. No guidance." This was an overstatement. There was always a plan of sorts, but if the word was not passed, perception was reality.

Iraqi Weapons of Mass Destruction

After the various task forces and investigative committees concluded that Saddam Hussein did not have any weapons of mass destruction in 2003, it was easy to forget the dimensions of this threat in the minds of the men and women who were about to go into battle. It was a threat they prepared for and lived with, in many cases for months. Every Marine, soldier, and airman was inoculated for anthrax and smallpox before coming to theater. Once in Kuwait, he carried a set of protective overalls in his backpack everywhere he went—along with rubber boots, gloves, and gasmask—and was prepared to use his atropine injectors to save himself when the seemingly inevitable attack came and he was "slimed." When on high alert, the Coalition forces wore the overalls over their uniforms in the desert heat and continued their mission. The best anyone could say about the heavy cloth protective gear was that it was not as hot as it seemed—and certainly not as hot as earlier generations of rubberized gear. Not only did the Coalition expect to encounter weapons of mass destruction when its forces reached Al Kut on the way to Baghdad, but it also feared that Saddam could launch strikes against troop concentrations, headquarters elements, or airfields in Kuwait, targets that were lucrative, close, and, in some cases, well known to the Iraqis who, after all, had occupied many of the same bases in 1990.

to replan continuously. I honestly believe [that] . . . as Americans we are going to do that anyway. We will plan for the last day. Once we begin executing, we will continue to plan. Our job is to make sure that we can take care of any contingency that comes up and give the CG . . . the game book of all the variations . . . to make things happen to win the campaign."[134]

There were various estimates as to the length of the war. While some of the early plans assumed a lengthy campaign, in the end the hope, and the expectation, was for a much shorter campaign. Two weeks if it goes well, two months if it does not, was what one of the Army generals told journalist Rick Atkinson.[135] In early March, General Conway commented that the three weeks predicted by the "talking heads," retired military officers who had been hired by the networks at home and were very free with their advice, was overly optimistic, because "Saddam has things he could do to slow us down."[136] After the war, General Conway's boss, General McKiernan, remembered that Phase III had been planned "as if there [would be] . . . determined fighting all along the way. . . . There were planning timelines that took it all the way [out to] 125 days."[137] One of McKiernan's planners, Major Evan Huelfer, remembered that General Franks sometimes sang a jingle that went "5-11-16-125," having to do with timelines: 5 days to position the final airbridge after the President made the decision to launch; 11 days to flow the final pieces of the "start force"; 16 days for the combined air and special operations attacks; and 125 days for the ground offensive.[138]

The uncertainty about timing matched the uncertainty about the enemy. Coalition Forces Command, and its major subordinate commands, seemed to be working off a set of assumptions that various agencies had developed for the contingency. General McKiernan explained those assumptions in an interview a few months later. The Iraqi forces were basically divided into three categories, the regular Army, the *Republican Guard*, and the *Special Republican Guard*. The Iraqi navy and air force no longer posed a threat to anyone but themselves, but there were approximately 21 ground divisions of various kinds and strengths for an estimated total of some 330,000 and 430,000 men. The regular Iraqi Army, which did not appear to be at a particularly high state of readiness, had up to six divisions in southern Iraq, including two that were relatively close to the border—the *51st Mechanized Division*, in and around the city of Az Zubayr near Basrah, and the *11th Infantry Division*, associated with the cities of An Nasiriyah and As Samawah on the border of the western desert. There were two other regular Army divisions in I MEF's area of operations that bore watching, the *10th Armored Division*, in the vicinity of Al Amarah, and the *6th Armored Division*, around Basrah.

General McKiernan surmised that Saddam Hussein probably intended his forces in the south to be no more than "a speed bump" for the Coalition. They would not put up much of a fight, and the Coalition's biggest problem would probably be what to do with all of the prisoners of war when they surrendered as they had in droves in Desert Storm. However, he expected that there would be something like a cordon around Baghdad of four elite *Republican Guard* divisions at a much higher state of readiness to defend the regime's center of gravity. At least two of these divisions, the *Baghdad* and the *Al Nida*, were likely to stand between I MEF and Baghdad. If and when Coalition forces approached the cordon, Saddam might attack them with weapons of mass destruction—that is, chemical or biological agents, which virtually everyone in uniform in the area of operations expected to encounter at some time during a war with Iraq. Inside the protective cordon around Baghdad there was also a "missile engagement zone," which meant the Iraqis were thought to have good antiaircraft defenses ranging from missiles to antiaircraft artillery to hand-held weapons.[139] The missile zone would be part of an urban Baghdad defense. The *Special Republican Guard*, estimated at around 15,000 soldiers and which was more like a palace guard than an army unit, would be part of that defense. Finally, there were the various irregular formations of paramilitary thugs like the Saddam *Fedayeen*. They were mentioned in most prewar assessments of the opposition, but they were seldom highlighted. The bottom line is that whoever the enemy turned out to be, no one thought the Baghdad fight would be easy.[140*]

Coalition Forces Command's assumptions were based in part on information collected by various sophisticated "national" means of collection. There was, for example, excellent overhead coverage; you could count the tanks in a tank park without any trouble. Task Force Tarawa's operations officer, Colonel Ronald J. Johnson, remembered he had voluminous order of battle information about the Iraqi Army, in some cases, down to the cell phone numbers of the Iraqi commanders, a truly impressive collection of data. What was lacking was hard information, especially about intentions. No one seemed to know what the enemy was thinking. Who could and would fight? What was the relationship between the regular army and the Ba'ath Party or the various special military and paramilitary organizations that the party had created?[141] In another example from the fall of 2002, the 1st Marine Division was exploring these questions and came up short. General Mattis wanted to organize an understudy program whereby his officers studied enemy division and corps commanders in order to understand them. But, given the shortage of human intelligence, "it was very hard, even for these dedicated young officers who were doing everything they could to . . . get information, to determine what was the background, what was the military school, what was their combat record, what [was] their political record," in short, to get a feel for their enemy based on the kind of information that military attaches traditionally collect.[142] As late as early March 2003, General Conway complained, in his diplomatic fashion, that there was "not as much intel coming in about Baghdad as we'd like" in order to plan that urban fight.[143*]

Not much more was known about Saddam Hussein's plans. Even after the war it took months for the Coalition to start to assemble a picture of his frame of mind, which appears to have lacked clarity and been unduly optimistic. He probably believed the Coalition would begin by waging a long air war, which would enable him to buy time for a cease-fire brokered by his friends in Moscow and Paris. Coalition planners generally believed that Saddam placed his hopes in various kinds of delaying tactics, both political and military. But according to a thoughtful article in *The Washington Post* in November 2003, investigators had been unable to find evidence of a coherent strategy for a ground war, such as a plan to abandon Baghdad and fight a guerrilla war, or to use weapons of mass destruction when Coalition forces penetrated too deeply into Iraq, despite interviews and interrogations of many former Iraqi officials and military leaders. A former Iraqi general who had been a division commander, Abed Mutlaq Jubouri, probably summed up the situation accurately when he said that Saddam Hussein failed to prepare his defenses in any kind of rational, systematic way. "There was no unity of command. There were five different armies . . . no cooperation or coordination. As to the defense of Baghdad, there was no plan."[144]

Planners throughout I MEF also wanted more information to drive their planning for Phase IV. There seemed to be even less processed and readily available information about the nature of Iraqi society than about the Iraqi Army—that is, information addressing the degree of popular support for the regime, or the challenges that the Coalition would

*U.S. Army estimates placed some 350,000 soldiers in the regular Iraqi Army and some 80,000 in the *Republican Guard*. (Rick Atkinson, *In the Company of Soldiers* [New York, NY: Henry Hold, 2004], p. 105)

*Historians Williamson Murray and Robert Scales reported that three U.S. Army generals commented that they were astounded by the depth of control the regime had over the people, and that another command was astounded by the tenacity of the individual paramilitary fighters. It was their contention that the various U.S. intelligence agencies should have been able to develop, highlight, and present this kind of information, much of which was freely available on the internet. (Atkinson, *In the Company of Soldiers*, p. 106; Reynolds, Journal, entry for 4Nov03) face in the wake of "catastrophic success," the shorthand for what was likely to happen if Saddam Hussein and his sons were overthrown by a sudden uprising. Another way to ask the question was what will Iraq be like after a successful offensive by the Coalition? What was the state of the infrastructure? How did the electricity work? How about the plumbing? What about the economy? Would the civil servants return to work and be able to do their jobs? All that existed in the way of answers to these questions were hazy assumptions that the Coalition might face a humanitarian crisis, for which it did prepare, especially by stockpiling food, and that it would be able to rely, at least to a certain extent, on Iraqi civil servants to get things going again.[145*]

*Some of this information existed inside CentCom and at other agencies inside the Washington Beltway. Gen Zinni had organized an interagency war game in 1999 to explore the challenges of rebuilding Iraq if Saddam Hussein were killed or deposed. For various reasons beyond the scope of this monograph, the topic was not a priority either for Pentagon or, by extension, CentCom planners at any time before March 2003.

Chapter 3

Preparing I MEF for War: The Most Important Fight is the First Fight

The major subordinate commands of I Marine Expeditionary Force did not wait for a formal order to prepare for war with Iraq. General Conway, who commanded 1st Marine Division before taking command of I MEF, had exercised the division's capabilities for war with Iraq in the first half of 2002.[146] When he took command from General Conway in August 2002, General Mattis put the division on a virtual war footing. From the outset his guidance was "to physically and mentally focus on one task . . . the defeat of the Iraqi Army and the liberation of the Iraqi people."[147] A few hours after the change of command, he gathered his commanders together in a secure room at I MEF headquarters at Camp Pendleton and told them to enjoy the coming weekend because it would be their last weekend off for a long time. They could take the time to ask forgiveness from the Almighty for what they would do to the Iraqi Army when the time came. In the meantime, he wanted them to train as if this week were the last week of peace.[148]

A general requirement for speed of execution and maneuver over distance governed individual and unit preparations. At many levels the division evaluated procedures, organization, and equipment with a view to being able to move and fight rapidly. For example, in September 2002 long-distance communications exercises took place, shifting control back and forth from the forward and the main command posts. Then the division structured its regimental combat teams to make them more robust and able to operate independently. The 5th Marines, for example, were reinforced with a battalion of tanks and a battalion of eight-wheeled light armored vehicles. Similarly, Mattis made sure there were enough tactical vehicles for everyone to ride to the fight and ordered modifications to enable the vehicles to carry extra fuel by welding "gypsy racks" to them, an initiative that did not find favor with the traditionalists in the motor-transport chain-of-command.

This was not simply a matter of general conditioning and preparedness. The division soon went beyond the general to anticipate specific missions, assigning them to subordinate units and staging elaborate rehearsals. This was especially true for what would become known as the "Opening Gambit," the division's scheme of maneuver for the opening days of the war. Around the division, there were mockups of various objectives in Iraq. At one point the parking lot in front of the division command post was taken over by a vast model of southern Iraq for an astoundingly detailed set of exercises, with toys usually used to create a make-believe world:

> The CG decided [that] using . . . Lego blocks to represent every vehicle in the division would be a fine way to visualize the . . . challenges . . . of moving massive numbers of vehicles down . . . the limited [number of] roads [in Iraq]. . . . Legos were available in a variety of sizes, and were color coded. Specific colors and sizes would be assigned to a unit's vehicles, then the blocks would be attached to the corresponding plates [!] [for] . . . each of the units of the division. A scaled terrain model was built to replicate the major terrain features in southern Iraq. . . . [E]ach battalion walked through its scheme of maneuver, moving [its] . . . Lego pieces in the proper sequence. . . . For each traffic jam of plastic blocks, the . . . audience was forced to ask itself, "Who owns the battle space? Where exactly are the boundaries?". . . These drills shaped the actions that would take place on the ground in Iraq. . . . For example, . . . the MAG-39 Operations Officer . . . [saw] that the AH-1 . . . [Cobras] were oriented to the east in support of RCT 7's attack on the 51st Mechanized Division . . . at dawn. . . . [T]he pilots would be attacking into the morning sun. The . . . plan was changed accordingly.[149]

Conway addressed the issue of supplying Marines stretched from Kuwait to Baghdad. In Enduring Freedom, a relatively small force, Task Force 58, had operated successfully some 400 miles from the beachhead. But now planners were talking about taking I MEF hundreds of miles into Iraq. By doctrine such a distance exceeded Marine Corps capabilities and spawned concerns that, logistically, I MEF was

JCCC 020125-N-6967M-506

Prior to assuming command of Marine Logistics Command, BGen Michael R. Lehnert, center, served as head of the joint task force charged with the custody of Al Qaeda detainees at Guantanamo Bay, Cuba. The organization would form the bridge between 1st Force Service Support Group up front and the theater-wide U.S. Army support command.

trying to go "a bridge too far." From the beginning, many senior officers, starting with the Commandant of the Marine Corps himself, focused on logistics as one of the three potential show-stoppers in Iraq, the other two being weapons of mass destruction and urban combat.[150] Luckily for the Marine Corps, inventive and energetic logisticians were ready to take on the challenge.

The I MEF subject matter expert for logistics was the commanding general of the 1st Force Service Support Group, Brigadier General Edward G. Usher III. General Usher knew the doctrine. The force service support group typically entered a theatre organized in stovepipes, that is, by function; there were headquarters and service, maintenance, supply, engineer, transportation support, medical, and dental battalions. Once overseas, the group tended to establish and maintain large stationary bases. If restructuring was necessary, it occurred in theater. Not flamboyant but thoroughly professional and forward thinking, General Usher decided he would task-organize to meet the challenge before leaving California, a process he started in the fall of 2001. The goal was to create combat service support units that integrated the various functions, something like the combined arms approach applied to combat service support to meet a variety of needs for the supported unit. For example, after Usher's reorganization there was Combat Service Support Group 11, which was to support 1st Marine Division through a combat service support battalion in general support, and three combat service support companies attached to the regimental combat teams in direct support. The battalion was expected to establish repair and replenishment points throughout the battlefield, while the companies followed in trace of the regiments, carrying one to two days of Class I (food), Class III (fuel), and Class V (ammunition) supplies. Another unit, Combat Service Support Battalion 22, was to support Task Force Tarawa, while Combat Service Support Group 13 supported the 3d Marine Aircraft Wing and its innovative plans to push supplies deep into Iraq. Providing general support to I MEF was Combat Service Support 15, required to have four days of supply on hand, as well as a full complement of transportation, engineer, and military police assets.[151]

For this war there were to be additional innovations by way of putting medical support virtually on the front lines, the shock trauma platoons and the forward resuscitative surgical systems would win praise for saving lives, as well as thoroughgoing preparations to conduct assault bridging operations, which were expected to play an important role in I MEF's area of operations. The ultimate goal of these preparations was better integration into the I MEF scheme of maneuver, from the initial stages of the planning process through the execution of the plan, and to enable I MEF to push the envelope a good bit further than it had ever been pushed before.

If 1st Force Service Support Group operated at the retail level, its wholesaler was to be the Marine Logistics Command (MLC). Comprising more than 4,000 Marines, mostly from 2d Force Service Support Group, the MLC was under the command of Brigadier General Michael R. Lehnert and remained under the operational control not of I MEF but of MarCent. It was intended to serve as a bridge between 1st Force Service Support Group and the theater-wide Army support command, the 377th Theater Support Command. The mission of the 377th was to support "Big Army" in a land war, as opposed to an expeditionary war, the Marine Corps' forte. But since the Marines were now preparing to fight a land war, they needed help from organizations like the 377th. General Hailston remembered an initiative to establish an MLC in Korea in the 1990s, and General Usher commented that it was an organization that had grown out of the lessons learned from the Gulf War. That is, the Marine Corps needed something it did not normally have—an operational logistics capability, at the echelon above I MEF. In his words: "the MLC basically [was] . . . our broker for overarching sustainment requirements for the theater."[152*]

Apart from such general statements of intent and a study by the Center for Naval Analyses, the Navy's think tank, there was little guidance and virtually no doctrine for General Lehnert to follow in preparing to stand up his command. The general was left basically to follow his own instincts and to learn through trial and error. He was a good choice to run a start-up operation, having recently weathered the challenge of standing up the task force to establish and run the detention center at Guantanamo Bay for the Al Qaeda detainees from Afghanistan. He had done so with just the right mix of common sense, good people skills, initiative, and energy. Now, in the Kuwaiti desert, he would need the same skill set. The MLC might look to the 377th for certain categories of supplies or equipment that were not in the Marines' inventory such as line haul (long-distance trucking) and heavy equipment transporters (usually used to transport tanks to save wear and tear). But if the 377th did not have enough assets to support both the Army and the Marines, General Lehnert would have to find work-arounds to obtain what the Army could not provide, in addition to maintaining wholesale stocks of Marine Corps supplies and ammunition.

Given the emphasis on speed and distance, it made sense for General Mattis to declare his intention to create the "most air-centric division in history" and to forge a close relationship with the 3d Marine Aircraft Wing under Major General James F. Amos. Amos, a former fighter pilot, had a good mix of command and staff time, including a tour as the head of the leadership section at The Basic School and combined/joint tours in Europe. His personal style was, by Marine Corps standards, laid-back; he was unusually approachable and looked for the commonsense solution as opposed to asserting his status or rank. That said, he was nothing if not results-oriented.[153]

What General Mattis said was certainly true, but it was not just the division-wing relationship that mattered; what was forged in 2002 was a very effective division-wing-force service support group team in the best tradition of the Marine air-ground task force. It was the personalities of the leaders, technology, and doctrine that came together to create an unusually powerful force. Precision weapons like the joint direct attack munitions, and changes in the doctrine for close air support, which made it easier to run, even

Photo courtesy of Col Charles J. Quilter II

A graduate of the University of Idaho and a naval aviator, MajGen James F. Amos held a variety of operational and staff assignments, including duty with the North Atlantic Treaty Organization, before assuming command of the 3d Marine Aircraft Wing in August 2002.

when no one could see the target, opened up a new range of options. The plan was for Marine air to take on its traditional missions, close air support, casualty evacuation, and occasional resupply, but Generals Conway and Mattis also wanted it to be a maneuver element in its own right, to take on more "independent" missions for the air-ground task force.[154]

This meant complicated discussions on what aircraft to bring to the fight and how best to use them. Through the process of global sourcing to augment its own resources, the wing put together an exceptionally robust team that would peak at 435 aircraft and some 15,000 Marines and sailors, making it the largest wing to deploy since Vietnam. The aircraft ranged from McDonnell Douglas FA-18 Hornet and McDonnell Douglas AV-8 Harrier fighter-bombers, to Lockheed KC-130 Hercules tanker/transports, to the same workhorses that Marines on the ground have, quite literally, been looking up at since the Vietnam War, the Bell UH-1H Huey and Boeing CH-46 Sea Knight helicopters. (Some older Marines remembered that the first CH-46s came on line in 1964 and have carried three generations of Marines into battle since then.) Two other helicopters, the Bell AH-1W Super

*By joint doctrine, each Service is responsible for supplying itself, especially with respect to Service-specific items. However, the combatant commander has the authority to create his own logistics structure, especially to designate theater support mechanisms, typically for "common-user" items.

Cobra, a remarkably versatile platform for close air support, and the Sikorsky CH-53 Sea Stallion, providing a heavy lift capability, rounded out the inventory.

Having more or less resolved the issue of who would control Marine air, CFACC or I MEF, in their favor, the force was largely free to use its air as it wished. The first priority was to gain and maintain air superiority, which would not take too much of the wing's time, given the state of the Iraqi air force and the drubbing that the Iraqi air defense system had taken in Operation Southern Watch, especially in the past few months. The next priority would be enemy command-and-control systems. At least initially, enemy artillery would rank almost as high as command and control on the list of priorities. This was in large part because everyone in the Coalition was worried about Saddam's ability to deliver weapons of mass destruction. (Any other targets associated with weapons of mass destruction that arose would also be a priority.)[155]

Many of these targets could be prosecuted as part of the "deep battle," deep being more of a concept than a location, although many deep targets would be far from what would pass for front lines in Iraq. The wing would be charged with reducing threats to I MEF before they could close with the Marines, or vice versa. General Conway wanted the wing to be prepared to take on Iraqi divisions that were still many road miles away from division's lead elements, and to reduce their effectiveness before they even started thinking about moving into battle. The *Republican Guard* divisions near Baghdad that were in the Marine area of operations were excellent candidates for everyone's list of deep targets. Closer in, but not necessarily that much closer in, General Mattis wanted the wing to defend his columns as they raced deep into Iraq, destroying enemy formations that could threaten the division's flanks and uncovering any other threats during the Marines' march up-country.

Sgt Carlos Carrasco of the 3d Light Armored Reconnaissance Battalion prepares to hand-launch a Dragon Eye interim-small unit remote scouting system at Camp Ripper, Kuwait.

DVIC DM-SD-05-05008

Since the battlefield would be fluid, without well-defined "friendly" and "enemy" lines, careful coordination would be necessary between the Marines in the air and on the ground. This called for some innovative thinking and organizing. The basic policy was to decentralize various functions and to keep them as close to the front as possible. This was true for "direct air support centers, air support elements, [and] imagery liaison cells from VMU [Marine Unmanned Aerial Vehicle] squadrons," flying unmanned aerial vehicles for reconnaissance, which became very popular with ground commanders, not to mention dedicated casualty evacuation and command-and-control aircraft.[156] Each regimental combat team would have its own dedicated aircraft and its own aviation support element. To enable this organization, the wing planned for a number of forward operating bases and especially forward arming and refueling points, the small mobile bases that would spring up alongside the advancing forces. In other words, for instance, if the commander wanted support from the wing, he would not have to work through some impersonal, centralized mechanism that would dispatch airframes from a base perhaps hundreds of miles away. Instead, he could call on the assets that were dedicated to him and that could be rearmed and resupplied locally.[157]

Typical of the intense preparations for air-ground cooperation was the rehearsal of concept (ROC) drill at the Wing Operations Center in Miramar, California, on 6 January 2003, with force and division commanders in attendance, along with all of the wing's group commanders. Intended, like the many other drills in Operation Iraqi Freedom, to refine the plan and make sure that all of the commanders understood it in the same way, this particular evolution was the first of its kind for a Marine air wing, which gives you a sense of how much more integrated the wing, already a major constituent part of the air-ground task force, would be in this fight.[158] The drill laid out the wing's plans for supporting both I MEF and the division; what was clear once again was the extent to which it would not only provide close air support to the ground combat element, but would also serve as a maneuver element in its own right as it attacked deep targets and moved its assets around the battlefield, both in the air and on the ground.

The "shared understanding" that emerged from the wing's ROC drill was one more example of that phenomenon in late 2002 and early 2003. General Mattis used those same words when he described the synergy that developed among Marine commanders at various levels, especially but not exclusively among the general officers who worked so well together under General Conway, whose command style was to welcome newcomers to the fold.[159] The commander of Task Force Tarawa, General Natonski, was happy to find General Conway ready to reach out to "the outsiders" from Camp Lejeune and make them "his own," narrowing the gap (which was as much perceived as real) between East Coast and West Coast Marines. "Shared understanding" continued to make it possible to work smoothly, both in the joint arena and within the air-ground task force, to rely on mission orders and the commander's intent and to operate with lean staffs and a "light" communications suite. The intended result was speed, the mantra that permeated I MEF, division, and wing planning, even before General Franks ordered his commanders to execute Operations Plan 1003V with the memorable words, "Make it fast and make it final."[160]

As the drills, and the planning, proceeded, Marines flowed into the country in great numbers. Apart from the Corps-wide "stop loss/stop move" decree, which applied to individuals, one of the most significant events in January was the approval of the deployment orders that began the wholesale flow of I MEF forces to Kuwait. According to I MEF's command chronology, the main deployment orders were 177A, issued on 2 January 2003, and 177B, issued on 14 January 2003. Together these two orders were ironically known as "son of the mother of all deployment orders," the Army's deployment order being the "mother of all deployment orders."[161] There has been considerable discussion of the use of deployment orders, with their ad hoc flavor, as opposed

JCCC 030215-N-2972R-065

Marines from the 2d Marine Expeditionary Brigade walk down the beach at a Kuwait Naval Base after disembarking from "Hopper 68," a U.S. Navy air cushioned landing craft.

to the more traditional, and many would argue, orderly, sequenced deployment data process.[162] The Pentagon decision to use deployment orders, apparently in search of strategic and political flexibility, did not meet with universal approval in the field. For one thing, it made planning that much more difficult, as each deployment, at least in theory, stood alone. Planners remember feeling that they needed to change the plan every time they generated a request for forces, which would go to the Office of the Secretary of Defense for approval and, with luck, turn into a deployment order. But the basic problem was still about sequence, as when a front line unit was separated from its enablers, that is, the units that supported it. A fair criticism from the Marine point of view was that the process could threaten the integrity of the air-ground task force, which was less likely under the deployment process.[163] Finally General McKiernan made what was perhaps the most telling criticism, from the warfighter's perspective, of the use of deployment orders. When asked about his reac-

Photo courtesy of CFLCC

Two Marines guide the driver of an M1A1 Abrams tank into a staging area after it was offloaded from a Maritime Prepositioning Force ship at a port in Kuwait. One Maritime Prepositioning Force squadron carried enough equipment, ranging from food and ammunition to tanks and howitzers, to outfit 17,000 Marines for 30 days.

tion to this "just-in-time" approach to delivering forces and equipment to theater, he replied with the very sensible observation: "I don't want them just in time. I want them a little bit early."[164]

Once they received their orders, I MEF units traveled to theater by sea and by air. There were the two amphibious task forces (ATF), known as ATF East and ATF West, which set sail in January and carried some 11,000 Marines and their equipment on 13 ships to Kuwait. Amphibious Task Force East carried the 2d Marine Expeditionary Brigade command element, and the elements that would serve under the brigade in Kuwait and Iraq, and arrived on 15 February.[165] Amphibious Task Force West carried Regimental Combat Team 1 (RCT 1), built around 1st Marines, and various aviation and combat service support units, arriving in late February.[166]

The remaining units came over by air, using a mix of military and chartered civilian aircraft. Some Marines literally flew first class, but for the majority it was a long, uncomfortable trip under crowded conditions. The 11 ships of the Maritime Prepositioning Force (MPF) squadrons moved independently to Kuwait with equipment for the 3d Marine Aircraft Wing and other I MEF units. Working 24-hours a day, seven days a week, Marines and sailors unloaded the ships in record time.[167] In the end, MarCent moved some 60,000 Marines and their equipment to Kuwait in less than 60 days, a staggering accomplishment. In December 2002 there had been only a handful of Marines in Kuwait, but by 1 March 2003, I MEF had about one-half of all the operating forces in the Marine Corps in country, facing north.[168]

Upon arrival in Kuwait, Marines and equipment moved into camps in the desert between Kuwait City and the border with Iraq that varied from relatively comfortable to very austere. I Marine Expeditionary Force had established its headquarters at Camp Commando, a few miles from Kuwait City. It had originally been acquired to serve as a no-frills, low-maintenance expeditionary camp for Marine units passing through Kuwait on routine exercises.[169] In the fall of 2002, however, as the force and division staffs trickled in, the camp had started its metamorphosis into a medium-sized military city. In February 2003 a member of the I MEF staff, Major Grant A. Williams, described Camp Commando in less than glowing terms:

> The area we obtained was an isolated portion of a Kuwaiti commando training facility. Last year we contracted to start pouring concrete slabs, [for] rudimentary plumbing, drainage, and [to] stockpile building materials. . . . Commando Camp is [now] a fortress carved out of the sand. . . . The physical layout . . . is approximately two miles around the perimeter . . . surrounded by concertina wire and fences. On the inside of the perimeter is a seven-foot berm that encircles the camp. This is an anti-RPG [rocket propelled grenade] . . . precaution that makes it difficult to get a direct shot into the camp. All the key points have guard towers with very focused young Marines manning them 24/7. In the center of the base is a seven-story tower the Kuwaitis use for repelling. . . . We have deployed a sniper team at the top of the tower. . . . With night vision goggles and infrared sights they can see and take out any bad Hadjis . . . a mile away day or night.[170]

Within the I MEF compound, surrounded by the tents of various sizes, stood three windowless "Butler" buildings made of sheet metal that looked like small warehouses. These became the combat operations and information center for I MEF Rear when the war started. Set up inside the nearly featureless buildings were cafeteria-style tables where staff officers with laptop computers controlled the force. One of the few decorator touches was a poster of *The Scream* by the gloomy Norwegian artist Edvard Munch, contributed by the operations officer, Colonel Larry K. Brown, Jr., who liked calling his domain the

Marine Order of Battle

By 0800 on 17 March 2003, the order of battle for I MEF, and the individual components strength and missions, were depicted in briefing charts at MarCent.

- I MEF Command Element–4,638
- 1st Marine Division–20,606–Secure the southern oil fields; conduct a passage of lines through Task Force Tarawa, and attack toward Baghdad.
- 3d Marine Aircraft Wing–14,381–Shape I MEF's battle space; screen the ground combat element from attacks; support CFACC.
- 1st Force Service Support Group–10,504–Provide direct combat service support to I MEF; interface with the Marine Logistics Command, a theater-level command under operational control of MarCent.
- I MEF Engineer Group–3,121–Maintain roads and bridges along the I MEF lines of communication; this unit was a composite of U.S. Navy construction battalions and Marine engineers.
- Task Force Tarawa (2d Marine Expeditionary Brigade)–5,091–Secure An Nasiriyah and crossings across the Euphrates River; secure lines of communication.
- 15th MEU–1,739–Attach to 1 (UK) Armored Division for Opening Gambit; attach to Task Force Tarawa.
- 1 (UK) Armored Division–21,045–Attack north from Kuwait; conduct relief in place in oil fields with 1st Marine Division; secure Basrah and vicinity.
- I MEF Total–81,125

Other Marine Forces in Theater:
MarCent Command Element (Bahrain)–385
Marine Logistics Command (Kuwait)–4,525
CJTF/Consequence Management (Kuwait)–742
MarCent Total–86,777*

* This is the rendition of the MarCent morning report, 17Mar03, captured by the field historian attached to MarCent, LtCol Jeffery Acosta, and sent to the author by e-mail. The total does not show the Marines committed to CJTF Horn of Africa.

"House of Angst."[171] There was also a mobile command post. Able to deploy to Iraq, I MEF Main was to be built around a structure known as "the Bug" because of its strange shape and because it bristled with antennae and radar dishes. Major Williams explained:

> We recently procured some high-speed, low-drag command tents that are similar to your basic self-erecting camping tent . . . on steroids. We can put up an interconnected dome style tent that can shelter 100 fully [functional] terminals hooked up to satellite feeds within four hours. Large . . . projection screens can display real-time satellite imagery and video feeds from unmanned aircraft. Environmental systems keep the inside cool (relatively). . . . [These are] not for the people but for the computers. [This command center is intended for service] well behind the front lines but [able] . . . to move forward so that the general can remain close to his battlefield commanders.[172*]

Apart from Camp Commando, there was quite a range of living conditions for Marines in Kuwait. Less austere, but still far from luxurious, were the Kuwaiti air force bases at Al Jaber, to the south of Kuwait City, and Ali Al Salem, to the west of the capital, where, thanks in part to other tenants like the U.S. Air Force, there were some creature comforts like a good, air-conditioned mess hall and shower trailers. But even at a "developed" base like Al Jaber, virtually

*"MEF Forward" was the term used when the Gen Conway left I MEF Main and went even farther forward with a very small staff. The Bug's footprint, just for its satellite dishes and related gear, was estimated at a mind-boggling one kilometer by one kilometer, and there were questions about why the I MEF even needed the Bug, since the communications suite in Kuwait was so good, as good as or better than the Bug's. The answer was that Gen Conway placed considerable emphasis on being physically close to his Marines. (Reynolds, Journal, entries for 10May03, 19July04)

DVIC DM-SD-05-03345

A below the horizon, aerial view of a 1st Marine Division unit command operation center and surroundings at an encampment in Kuwait. By mid-March 2003, the area would house more than 20,000 Marines.

all Marines lived and worked under canvas. When you drove onto the base at Al Jaber, you passed under an archway where it was said Saddam Hussein's forces had hanged Kuwaiti air force officers when they took over the country in 1990, just to let everyone know who was now in charge, a potent reminder of why the Coalition had returned in 2003.

Some Marines lived and worked at Camp Doha, a sprawling U.S. Army base between Camp Commando and Kuwait City that looked like a prison in

Catching the Bug

The "Bug" was the brainchild of I MEF's intense chief of staff, Colonel John C. Coleman. In the fall of 2002 he sold the concept to General Conway and then went to find the right tentmaker, whom he located in Virginia. Sitting on the floor of the tentmaker's shop, he had sketched his concept. A few days later the finished product appeared at I MEF headquarters in California. The force took it to Kuwait, and tested it in the desert outside Camp Commando, passing command back and forth with I MEF Rear. It offered the commander "an incredibly rich picture of the battlefield." Screens in the command center could display a mind-numbing array of data, from live Cable News Network reports from the front lines, to customized maps, to satellite imagery. The software even allowed the commander to "test drive" potential routes through built-up areas. Colonel Coleman said that putting the "Bug" together was the easy part. The hard part was educating the staff, which he divided into the "Flintstones," the stone age warriors, and the "Jetsons," the space-age warriors. But Flintstones could grow into Jetsons only when they had the right equipment. The "Bug" replaced the old maps covered with clear acetate, overwritten with grease pencil, and updated with data from old-fashioned line-of-sight tactical radio nets. That is what many subordinate Marine units still had, where there were only a few technological marvels like the "Blue Force Tracker," a laptop computer screen with ground-satellite links showing the location of friendly units and even allowing their commanders to exchange a few words.*

*Col John C. Coleman intvw, 11Dec03 (MCHC, Quantico, VA); David J. Lynch, "Marines' Mobile War Room Is Rich with Data," *USA Today*, 31Mar03, p. A-06.

DVIC DM-SD-04-00296

Originally a barren patch of desert in northern Kuwait, Camp Coyote rapidly became a massive logistics hub for 1st Service Support Group Forward and the headquarters for Regimental Combat Team 7.

the desert, with its watchtowers and rings of security provided by U.S. contractors. The guards were retired U.S. military personnel making good tax-exempt money. Doha was home for the Marines who worked at C/JTF-Consequence Management and for the Marines on the Coalition Forces Land Component Command (CFLCC) staff. The Coalition Command had an incredibly sophisticated command and information center, much more elaborate than the "Bug," and some of the accommodations at Doha were among the best in theater. The camp even boasted some semi-private, air-conditioned trailers. But much of the accommodations were far more basic, like the warehouses where soldiers and Marines simply set up their cots in large, open bays in the withering heat.

Apart from Camp Commando, Doha, and the air bases, most of the camps in Kuwait were simply patches of desert where Marines pitched their tents and bulldozed sand to form a protective berm around the perimeter. General Mattis had said he wanted everyone in his division ready to live like an infantry lance corporal in the field, and, for the most part, circumstances obliged him. The division's Camp Matilda was in the middle of nowhere and comprised a number of large 50-man tents, mostly rented from the Kuwaitis. They looked like the U.S. military's general-purpose tents but were much larger, designed for traditional Arab social gatherings in the desert. Now they were surrounded by oceans of equipment, mostly from the MPF ships, much of it still painted green or "Woodland" camouflage, and by two-man tents pitched here and there wherever there was a vacant bit of sand. No one knew for sure exactly how big these encampments were, but they were not small. For example, the commander of Regimental Combat Team 7, Colonel Steven A. Hummer, gauged the size of his regiment's encampment within Camp Matilda by the amount of time it took him to jog around it, 30 minutes.[173]

No matter which camp they called home, most Marines in Kuwait had to put up with frequent sandstorms that could blind a man as surely as a blizzard in the Dakotas, as well as temperatures that could sink to near freezing at night and soar to well over 100 degrees Fahrenheit during the day. Every Marine, soldier, and sailor in Kuwait knew he was within the "Scud fan," that Saddam Hussein could fire missiles at the American forces that were concentrated in a small area. Many feared that Saddam would load the missiles with chemical weapons to "slime" his victims.* To guard against that possibility, U.S. Army Patriot antimissile missile batteries were on the ready throughout Kuwait. Some Patriot batteries from the 108th Air Defense Artillery Brigade were under I MEF's tactical control and were slated to move with the Marines into Iraq. The Army was, by and large, generous with its attachments, sending some 2,700 soldiers to I MEF with specialized talents that the force needed. Most places also had crude bomb or missile shelters, which were often simply inverted, U-shaped concrete culverts, about three-feet high that would have provided some protection from a blast

*This was another bit of jargon peculiar to the Iraq War, often seen as a passive verb, as in "I know I am going to be slimed." There was not a lot of clarity about how this would happen, but one bit of scuttlebutt was that there could be an airburst that would shower contaminants on everyone within a grid square.

but almost none from a chemical strike. Anyone doing the math could see there were not enough spaces in the shelters for everyone wearing a uniform in Kuwait.

If a chemical strike had occurred, each unit's organic decontamination teams would have sprung into action, helping to decontaminate people and things, which would have required a great deal of water and a lot of time. There was also an expectation that the Marines and some of the foreign experts in Combined Joint Task Force Consequence Management (C/JTF-CM) would pitch in. This was true even though the largely German and Czechoslovak international force was in Kuwait as part of Operation Enduring Freedom, not to participate in a war with Iraq but to "be prepared to" assist host nations in the event of a terrorist attack with chemical or biological agents. The combined-joint task force still answered to I MEF, and its predominantly Marine command element, commanded in 2003 by Brigadier General Cornell A. Wilson, routinely participated in CentCom planning conferences.[174]

Marines with Headquarters Company, 5th Marines, practice changing into mission-oriented protective posture response level 4 (MOPP-4) suits during a nuclear, biological, chemical drill at Camp Coyote, Kuwait. Because of the near-unanimous belief that Saddam Hussein would use chemical weapons, the frequent alerts were taken seriously.

DVIC DM-SD-05-03779

After the Marines fell in on their gear, there was the process to test and calibrate equipment and to hone combat skills. I Marine Expeditionary Force coordinated more planning and more drills, all of which were remarkable for their thoroughness. War games were carried out to explore the branches of the plan, the possible nightmares like "early military collapse" and intentional flooding. On 7 February, division hosted a major rehearsal of concept drill at its Camp Matilda for the force on a vast "model of the obstacle system of northern Kuwait, [and] the oil fields, the rivers, the ports, the roads [on the Iraqi side of the border]."[175] Framed on one side by four seven-ton trucks forming a kind of grandstand overlooking a piece of ground the size of a football field, the exercise went for three hours, with U.S. Marines, Royal Marines, and British soldiers, all in colored jerseys representing their units, literally walking through the moves that would occur during the first few days of the war. Major General Stalder, now deputy commander of I MEF, spoke about the utility of rehearsals and the need to form an image of the plan. This was already a familiar refrain, that Marines at all levels not only understand but also visualize the plan, "seeing" in their mind's eye the terrain, the enemy on that terrain, and their own actions in battle. As General Mattis had put it in a memorandum on 20 December 2002: "It is critical that each of us anticipate . . . what lies ahead and continue . . . to visualize (or image) our troops through the challenges . . . so well that they will move through [them] with a sense of déjà vu."[176] As he had done before, Mattis now warned against becoming wedded to the plan, reminding his listeners that once I MEF crossed the Tigris River, they would have many options for getting to East Baghdad and lots of room for individual initiative and action. This was at a point when the plan for the Baghdad fight was still very much up in the air, although it was generally assumed by now that the Marines would approach Baghdad from the east, and, most likely, enter the city itself.* Nevertheless, Mattis also made the point that the carefully choreographed plan would give his Marines confidence in the first days of the war, which would be the baptism of fire

*The command cruise book would state that during this period the division was "careful not to be wedded to a base plan. The tactical, physical, and moral readiness of the individual Marine was to . . . determin[e] . . . success or failure, not reliance on a scripted plan." (Capt Lara A. Bennett, et al., *No Better Friend, No Worse Enemy* [Camp Pendleton, CA: 1st Marine Division, 2004], p. 13)

Consequence Management

Morning formation at the Combined Joint Task Force Consequence Management (C/JTF-CM) at Camp Doha was an unusual sight—the largest body of troops was German. Seeing and hearing German spoken by men in uniform in the desert, anyone who had ever studied World War II could not help but think of another time, especially when next to the formation he could see the iron crosses painted on a tan background on the Germans' elaborate "Fox" vehicles. These vehicles, which looked like light armored vehicles to Marines, carried state-of-the-art technical gear to sample and sniff for NBC agents. The force had a few Fox vehicles of its own that would travel north when the time came. What was very different this time was that standing in the formations next to the Germans were Czech and Slovak soldiers, many of them women, with similar missions. The idea was that in the event of an attack, the Germans would "chase the plume," following the fallout and figuring out what the agents were, while the Czechs and the Slovaks would concentrate on decontamination and on "turning victims into patients," meaning they would conduct triage and start medical treatment if there were mass casualties. Every so often, C/JTF-CM would conduct a field exercise. One such exercise occurred on 8 April 2003 in Tactical Assembly Area Fox, the home of the Marine Logistics Command in the gently rolling Kuwaiti desert, otherwise featureless but for a few tufts of grass. After drawing weapons and ammunition, a routine force protection measure even in Kuwait in April, the group set out from Camp Doha and bounced around in tactical vehicles for about 60 to 90 minutes, through the outskirts of Kuwait City, which looked like one big auto salvage yard (some of the material was "Iraqi surplus" left over from Desert Storm) and into the desert, from good highway to rough paved road, to gravel road, to desert track. Near the MLC's headquarters, there was an exercise command post set up in a general-purpose tent without air conditioning. It was a wonder that the computers worked, since the temperature was in the low 100s Fahrenheit and rising. There had been a communications glitch, and the Germans had been held up at the gate to the Marine Logistics Command. Visibly annoyed, their commander took the astounding step of declaring that since the exercise was unrealistic, he was exercising his prerogative to cut short German participation and would return with his unit to Doha. The Czech contingent, however, turned to with redoubled enthusiasm and proceeded to practice decontaminating vehicles. The decontamination site looked like a car wash in the desert, except that the workers were working hard in completely sealed Soviet-style rubber suits in the staggering mid-afternoon heat. The Czech brigadier, who was cheerful, realistic, and easy to talk to, was in the thick of the action, an officer in his element.*

Photo courtesy of Field History Branch

Gathered around a map of Kuwait, Kuwait City fire chief LtCol Manei Al-Hayan and members of the Combined/Joint Task Force-Consequence Management: operations chief GySgt Osama B. Shofani, senior German nuclear, biological and chemical officer Maj Andreas Kayser, and initial response commander Marine LtCol Charles G. Chase, plot the location of a downed Iraqi missile in preparation for deploying forces if there are chemical agents present.

*Reynolds, Journal, entry for 8Apr03.

Photo courtesy of Col Charles J. Quilter II

The principal leaders under I Marine Expeditionary Force were: first row from left to right, BGen Edward G. Usher III, commanding general of 1st Force Service Support Group; LtGen James T. Conway, Commanding General, I Marine Expeditionary Force; and MajGen Robin V. Brims, 1 (UK) Armored Division commander; second row, MajGen James N. Mattis, Commanding General, 1st Marine Division; MajGen James F. Amos, Commanding General, 3d Marine Aircraft Wing; third row, MajGen Keith J. Stalder, Deputy Commanding General, I Marine Expeditionary Force; and, BGen Richard F. Natonski, Commanding General, 2d Marine Expeditionary Brigade (Task Force Tarawa).

for many of them. He predicted that the force would have about 30 days more to prepare for war. The field historian who recorded the evolution noted that the mood was "very somber . . . no ooorahs, almost no laughter."[177]

The next day, I MEF participated in CFLCC's four-day exercise "Lucky Warrior 03-2," intended to test command and control for the two-corps war plan that the CFLCC commander, General McKiernan, had called for after the last round of exercises in November and December. The exercise was labeled a "dress rehearsal" for war; the CFLCC Operations Plan had been published on 13 January, which meant I MEF could finalize its plan, ultimately published on 10 February.[178]

Hard on the heels of Lucky Warrior came the CFLCC rehearsal of concept drill around the only slightly smaller terrain model in one of the warehouses at Camp Doha on 14-15 February. It integrated the results of previous exercises and gave CFLCC components an opportunity to appreciate how their units would fit into the overall plan. Within the next month, there were at least two other major rehearsal drills, one at the division level on 27 February, and one at I MEF level on 10 March, not to mention similar evolutions at Task Force Tarawa, all serving to imprint the plan on the minds of the people who would execute it. At the 10 March evolution, General Conway made the point about the value of ROC drills in his terms, saying that the "most important fight is the first one," the idea being that if the force could win the first battle, each succeeding battle would be that much easier.[179]*

The rehearsal of concept drill on 27 February was particularly memorable, a capstone event of sorts. Division engineers again prepared the ground with D-7 bulldozers, constructing on the desert floor a multi-tiered amphitheater, about 100 meters in length, with an angled surface for better viewing from the cheap seats. The audience included Generals Conway, Amos, Natonski, and Usher, as well as Major General Robin V. Brims, the British division commander. Once again, Marines and soldiers in colored jerseys on the "board" stepped through the actions that their units would take after crossing the line of departure. Division hosted the event, with the stated purpose of putting its intentions, and interac-

*In the run-up to combat operations, subordinate commanders down to the division level would use the terrain model at Doha again to brief Gen McKiernan and his staff on their plans.

tions with other units, on display. General Mattis repeated the by now familiar refrain: all of the "players" had to be able to visualize the "battlefield geometry" and the sequence of events in the first few days of the war, not unlike athletes who are trained to visualize what they are going to do on the playing field. He called it "anticipatory decision making," adding "anticipation to be one of the most useful abilities of a field grade officer . . . [meaning the ability] to anticipate . . . both friendly requirements . . . and what the enemy's going to do."[180]

The commanders used the rehearsal drills, and other assemblies, to deliver a series of "go to war" speeches, speaking to their troops about the coming challenges in a more personal way, and continuing to erode the doubts among some that there would even be a war, not an unrealistic response to all of the clamor against the war throughout the world and the calls for a negotiated settlement of some sort.*

General Conway set out to visit as many units as possible and to deliver his message in person. On 1 March, he went to nearby Camp Ryan to appear before Task Force Tarawa. Tarawa's headquarters group, and all of Regimental Combat Team 2, formed a box of companies in the desert. Armored vehicles formed one side of the box, and General Conway mounted the M1A1 Abrams tank in the center of the side to give his talk, making his points over the high wind and dust.

The general had three main themes, why the force was there, what it was going to do, and the individual Marine in battle. The mission was to forestall future terrorist attacks like that of 11 September. Either Saddam would disarm, or the Coalition would go to war when President Bush gave the word. When and if that happened, I MEF would fight three battles: the deep fight, from the air, to reduce the enemy's combat effectiveness by up to 50 percent; the close fight on the ground; and the rear battle, that is, providing sustainment. The Marines had two things going for them, their reputation as the "meanest in the valley" had preceded them, it must have had an effect on their enemy, and, of course, their combined arms approach. Marines on the ground and in the air fought as one. At this point, on cue, there was a flyover by Cobra attack helicopters and jet fighters. When the noise of the flyover died down, General Conway reminded his Marines that they were not there to fight

DVIC DM-SD-04-11709

Standing atop an M1A1 Abrams battle tank, LtGen James T. Conway addresses the Marines of Regimental Combat Team 7 at Camp Coyote, Kuwait. In these talks he covered why Marines were there, what they would do, and the role of the individual Marine in battle.

Iraq, but rather Saddam's regime and they needed to distinguish between those who wanted to fight and those who did not. Finally, speaking to the individuals in front of him, Conway said that the next few days would govern how they saw themselves, and were seen, for the rest of their lives. Fear was a "natural battlefield phenomenon" that could sharpen perceptions and make Marines react more quickly. The Marines had better gear, weapons, and health care, but at the end of the day the battle was about people, and our people were better. He told the Marines to take care of one another, and themselves, and invoked a blessing before ending his talk.[181]

After addressing the formation, Conway gathered the officers of the task force together and dwelt on

*Col Dennis Judge, I MEF's current operations officer, stated he did not feel sure there would be a shooting war until the beginning of March, unlike Gen Mattis, who seemed to have "known" all along that there would be war. (Col Dennis Judge intvw, 11Aug03 [MCHC, Quantico, VA])

some of the same themes in more detail. One theme was the protests against the Iraq war, both at home and abroad, that looked like Vietnam-era antiwar protests on television. He suggested that Tarawa's leaders talk about the protests with their troops, making the points that this was not a war about oil and that his Marines would not be treated like Vietnam veterans. They would be honored when they returned home. He expressed his belief that in the near future the President would issue an ultimatum and that when it expired the force would attack. The Marines needed to remain flexible; the plan was always changing, there was no guarantee that anyone would draw the same mission he had planned for. He said that while Tarawa's initial role was to be in support that could change, I MEF may need to reorganize and "retask," say, to head to the oil fields north of Baghdad.

Conway stressed the standard rules of engagement, declaring that they were not as restrictive as they might have been. On the other hand, Marines had to be careful—the mission was to unseat a dictator, not necessarily to destroy his army, and certainly not to kill civilians. The four specific points he made were:

(1) Commanders always have the right to self-defense.

(2) An enemy commander who purposefully places his forces near civilians has violated the law of land warfare.

(3) A commander is responsible for his actions based on the facts as they appear to him at the time, not as they will appear in retrospect. It boiled down to a matter of good judgment.

(4) Use "Wilhelm's Law."* If the enemy fires first, the Marine response should be proportional, use the smallest weapon first, and then progress upward. If Marines initiate fires, they need to mitigate "collateral damage" to individual civilians and civilian buildings.

Conway went on to express his concerns about fratricide; Marine weapons were more lethal than the enemy's, and they needed to be sure of their targets before they fired. No one wanted to live with the responsibility of having killed friendly troops. Next he returned to the subject of fear. He wanted Tarawa's officers to talk about it with their troops. It was a problem that each would solve in his or her own way, some thinking they were invincible, others, like Confederate General Stonewall Jackson, believing that all was in God's hands. That said, General Conway told his officers that their greatest fear should be that they let their fellow Marines down in some way, by choking or by being crazy-brave. He cited the example of a second lieutenant in Vietnam who wanted to go home with a medal and was last seen charging the enemy with his .45-caliber pistol at high port. He did nothing for himself, his Marines, or his family.[182]

General Conway's speech varied slightly as he spoke both to other large formations of Marines and to smaller groups of officers, who were invited to ask questions. Generally, however, his remarks were brief, motivating, and practical. There were, for example, admonitions to leaders not to forget to sleep during battle, and there were moments of humor, as when he quipped that "when Abdul in the 51st Mechanized Division north of the border heard that he was taking on the 1st Marine Division followed by the 1st UK Division, he said . . . , 'Ana felaka beluchi,' which is Arabic for 'Ain't that a bitch!'"[183]

On 14 March, Conway made his way over to the British encampment in the desert a few miles away for what he considered a most impressive welcome and what some British officers considered one of the most inspiring speeches they had ever heard.[184] To reach the formation of thousands of British troops of the 7 Armored and 16 Air Assault Brigades, he drove between two lines of Warrior armored personnel carriers, whose crews saluted him as he passed. Finally he reached the heart of the formation, dominated by two enormous cranes or mobile bridges that formed the letter "M," ostensibly for "Marine," where he was welcomed by pipers from the Black Watch. Then, using a loudspeaker that, the general was sure, could be heard in Iraq, the British sergeant major called the formation to attention and presented it to him.

Standing on top of a Challenger II tank decorated with a British Union Jack and the Stars and Stripes, Conway told the British that it was great to have them aboard. "Two hundred and fifty years ago the expression 'The British are coming' would scare [American] . . . children." But in Kuwait in December and January, that expression had been "a very positive thing." As he had in other speeches, he spoke about the reasons for war and about the strength of the Marine air-ground team. He said that one of four things would happen: Saddam could turn over his weapons of mass destruction; he could leave the country; "someone up there might kill him; [and] . . . if that does not happen, we are going to kill him." At the appropriate moment, Cobras and Hornets flew by, the Hornets blasting "into view in a split second" at 200 feet before fanning out into the clear blue sky and disappearing from view. Speaking of the British

*The law was named after retired Marine Gen Charles E. Wilhelm, who had come to theater as a mentor in December to observe I MEF's preparations for war and offer his advice.

troops' fighting prowess, Conway borrowed a quote attributed to George Orwell: "Our countrymen should rest well in their beds at night because rough men stand ready to visit violence on those who would do them harm."[185]

Subordinate Marine commanders echoed many of the same themes as they made the rounds and addressed their troops. Like General Conway, they spoke about honor and values in a way that was unfamiliar to many Cold War Marines. General Mattis had been preparing his troops for the inevitable for some time, most notably during the rehearsal of concept drills in late February. On 27 February, he had spoken forcefully about "soldierly compassion and incredible violence."[186] One of the officers at the rehearsal drill, field historian Colonel Reed R. Bonadonna, wrote what he had heard:

> General Mattis was an inspiring and sometimes fiery speaker, and he did not disappoint. . . . [H]e provided the words of inspiration to pass down to the troops. He said that the Brits and we were free people fighting for what we believe in. There are no war protests in Iraq . . . because Saddam would not allow them. He told us that, when we get home, we should shake the hand of a war protester and thank him for exercising the freedoms we had fought to protect, and then wink at the protester's girl. . . . [This] brought the house down, releasing the tension so many of us felt . . . Mattis . . . appeal[ed] . . . to our sense of justice and our sense of ourselves as men in the same breath. . . . He was like Patton, maybe better, because [he] . . . came across as natural, not performative.[187]

On one of his many visits to the units under his command, MajGen James N. Mattis stressed that the 1st Marine Division would destroy those Iraqi forces that chose to fight, and treat all others with decency and soldierly compassion.

CCC 030218-M-5150A-006

In a written message in March, General Mattis summarized the reasons for the war and declared that "together we will . . . close with those forces that choose to fight, and destroy them. . . . [W]e will treat all others with decency, demonstrating chivalry and soldierly compassion." He warned his Marines to be ready for "chemical attack, treachery, and . . . other unethical tactics," exhorting them to keep their honor clean and to keep faith with their comrades on their left and right and with Marine air overhead. Finally, in an archaic turn of phrase, he charged them to "[f]ight with a happy heart and strong spirit."[189] The general's intent was to evoke Sitting Bull and Crazy Horse on the Little Bighorn, not General Custer, who may have had a strong spirit but probably did not have a happy heart on the day he died. It was "the idea that . . . you sense that all is well in your world [as] . . . a Marine . . . you have got a good Marine on your left and [on] your right. . . . [T]hat sense of happiness [about] . . . who you are fighting alongside armors you . . . against the trauma of the battlefield."[189]

Like General Mattis, General Amos had been making the rounds of his squadrons. The gatherings on 4 March on the USS *Bataan* (LHD 5), now the "Harrier-carrier," were typical. He spoke first to the assembled Marines and sailors in the well deck and then met with the officers of the two embarked squadrons in the ready room. The general implied at each meeting that the war would start soon. He emphasized a number of themes: watching out for one another; the just cause of the Coalition's impending actions, to be taken against Saddam Hussein, not the Iraqi people; the corollary need to limit collateral damage; and the belief that many Iraqi soldiers would surrender rather than fight.[190] Just before launching his forces against the enemy, General Amos wrote a more formal message to the wing, offering a "few thoughts" to his Marines about the honorable cause that they were embarking on and declaring that there was "a fear worse than death . . . that is the fear of letting down your fellow Marines." He predicted, "We will win this war and the respect of the Iraqi people . . . and we will do it honorably." He distributed the message to his air groups as an attachment to an e-mail, which he ended with "God Bless each of you . . . now let us go kick the shit out of them!"[191]*

Perhaps the most eloquent "go to war" speech that anyone delivered was that of Lieutenant Colonel Tim Collins, commanding officer of the 1st Battalion, Royal Irish Regiment, delivered on 19 March. Known for his American-made gold Ray-Ban sunglasses, ever-present cigar, and kukri, the ferocious curved Nepalese blade he was entitled to carry for commanding the Gurkha company attached to his battalion, Collins was a flamboyant commander whose ancestors had served with the regiment for generations.[192] Said to be a reflection of his upbringing and his professionalism, and not of a desire for publicity, his unusual eloquence was widely reported in the press and caught the attention of a number of prominent figures, including Prince Charles and President Bush:[193]**

> We go to liberate, not to conquer. We will not fly our flags in their country. . . . There are some who are alive at this moment who will not be alive shortly. Those who do not wish to go on that journey, we will not send. As for the others, I expect you to rock their world. Wipe them out if that is what they choose. But if you are ferocious in battle, remember to be magnanimous in victory. . . . If there are casualties of war then remember that, when they woke up and got dressed in the morning, they did not plan to die this day. Allow them dignity in death. The enemy should be in no doubt that we are his nemesis and that we are bringing about his rightful destruction. . . . You will be shunned unless your conduct is of the highest . . . We will bring shame on neither our uniform nor our nation.[194]

The speeches, along with the frequent air-raid or, more precisely, missile-raid drills, to prepare for a possible preemptive strike with weapons of mass destruction against the troop concentrations in Kuwait, all contributed to a sense that war was imminent. There was also a sense among some officers that the operation was a high-risk proposition on at least one other account, and that was the logistics challenge: Would I MEF be able to push supplies fast enough into the heart of Iraq to maintain the momentum of the attack?[195] But along with these concerns was a wish for the waiting to end and a determination to get on with the inevitable while the Marines were at their peak. If left too long in their desert camps, these young men and women might lose their edge. "Waiting [was]," a field historian wrote, "hard on morale . . . from [the] youngest to [the] most senior."[196]

*BGen Richard Natonski of Task Force Tarawa and the 1st Force Service Support Group's deputy commander and sergeant major also made the rounds of their subordinate units, making final coordination and delivering words of encouragement.

**One of the ironies of the war is that two months later the press also reported that Collins was under investigation for mistreating prisoners. Maj Re Biastre, a U.S. Army civil affairs officer attached to Collins's battalion, lodged a formal complaint against Collins that contained a number of charges. It soon emerged that on the streets of occupied Rumaylah, Biastre had challenged Collins's authority in the presence of Brigadier Jacko Page, Collins's immediate superior. Collins had reacted angrily, ordering Biastre's arrest for insubordination and then banishing him from the area. Biastre in turn prepared a 2,400-word statement including specific charges against Collins along with the observation that many British officers had expressed their resentment of Americans. The charges were mistakenly attributed to Maj Stanton S. Coerr, an ANGLICO Marine who was also attached to the battalion but outraged by the mistake, since he held Collins in high regard. In the end, after considerable press play in the United Kingdom, the Ministry of Defence announced that the charges had been dropped. The incident became a footnote to history that sheds light on the nature of British-American relations during the war. Biastre's complaint was not entirely misplaced. Occasionally undercurrents of tension between the United States and the British in Iraq would surface, which was natural since the British public was opposed to the war and since the culture of the long-serving British military was so different from that of the short-serving American military. Some British officers appeared more likely to stand on their authority than their less formal American counterparts. But on the whole, the United States Marines and the British soldiers and Royal Marines worked together with less friction than American sister Services in past conflicts, and happily drank together at the end of Phase III. ("Ministry Clears RIR Iraq Hero," *Belfast News Letter*, 2Sep03, p. 6)

Chapter 4

The Opening Gambit: "Tally-Ho!"

The "Opening Gambit" that I MEF prepared so thoroughly in its drills and plans still had as its goal to seize the southern oil fields that were, very roughly, north of Kuwait and west of the city of Basrah, and usually known as the Rumaylah oil fields. For General McKiernan, seizing the oil fields was like a foreign branch to his base plan, separate from the essential mission of driving to Baghdad, considered the enemy center of gravity.[197] It was an imperative, imposed on Coalition Forces Land Component Command (CFLCC) by CentCom rather late in the day, intended to keep Saddam Hussein from creating an environmental disaster by blowing oil wells as he had in Kuwait in 1991 and as he planned to do again in his own country when the Coalition attacked. The way General Conway explained it, for CentCom the oil fields were as important as Baghdad, and money from oil was needed to rebuild the country.[198]

The final plan was for the division to seize the oil fields while the British, reinforced by 15th Marine Expeditionary Unit (Special Operations Capable) (15th MEU (SOC)) to make up for a British unit that could not make it to the fight, moved against the Al Faw Peninsula and the port of Umm Qasr, which lay between Kuwait and Basrah. Once Umm Qasr and the oil fields had been secured by the Marines, the British Army would relieve them and become responsible for securing the southeastern part of Iraq. The 15th MEU would revert to Marine control, and I MEF would move to the west, on its way to the river crossings at An Nasiriyah before moving north.

The plan for the first few days of war was so carefully choreographed, as contrasted with the more general plans for the rest of the war. Even though the nature of Marine participation in the Baghdad fight was still up in the air, the Marine expectation, as of early March, was still that the force would pass through Al Kut to threaten Baghdad from the east.

Within 24 hours of the 17 March movement order, the 1st Marine Division's 20,000 Marines and more than 5,000 vehicles deployed from support areas into dispersal areas. The shift began a series of moves that would take the division to Baghdad.

Photo courtesy of 1st Marine Division

Photo courtesy of CFLCC

LtGen David A. McKiernan, right, commander of the Coalition Forces Land Component Command, talks on the phone in the "War Room" at Camp Doha, Kuwait, to his subordinate commanders as combat now looked inevitable.

General Conway considered the danger of attack by weapons of mass destruction to be "highest" in and around Al Kut; it would be on the route to Baghdad but not too close to that city (the theory being that the enemy would hesitate to contaminate his own capital). That militated for getting the Marines as close to Baghdad as they could and as quickly as possible, whatever the final arrangements for the difficult fight inside the city turned out to be.[199] Nevertheless, General Mattis made it a matter of record that he was ready not only to cross the line of departure but to go on to the capital: "this division is prepared for rapid attack in the Iraq regime's center of gravity, Baghdad."[200]

On 17 March, President George W. Bush gave Saddam Hussein and his sons 48 hours to leave Iraq. What followed was a confusing series of schedule changes for I MEF. Now on edge about sabotage in the oil fields, CFLCC sent a "be prepared to" tasker to Conway. This led I MEF, in the early morning hours of 18 March, to issue Fragmentary Order (FragO) 046-03, which tasked its subordinate commands to be prepared, by 1800Z (or 1800 Greenwich Mean Time, a standard used to avoid confusion) on the same day, to seize the oil fields on four hours' notice. The next day, after seeing "live-feed" from a "Predator" unmanned aerial vehicle showing oil well fires that looked like sabotage, CFLCC contacted CentCom to request permission to launch the ground offensive early to in order to limit the potential for further sabotage.[201]

Long lines of Marine vehicles now started to move to their dispersal areas and then to their intermediate attack positions near the border. The processions moved across the desert landscape on 18-19 March, while engineers finalized the complicated work of clearing lanes through the demilitarized zone on the border between Kuwait and Iraq. Although often called "the berm," as in, "I am going across the berm into Iraq," it was actually much more than that. In most places there was at least one antitank ditch, a 10-foot berm, and an electric fence. With Kuwaiti assistance, Coalition engineers had been working on the berm for quite some time to prepare lanes for the attack.[202] The adrenaline was starting to flow. Ground crews and aircrews turned to at the large bases like Ali Al Salem and Al Jaber, as they did at a few small expeditionary airfields like "Joe Foss," which was little more than a rolled-sand landing strip for KC-130s in the desert near the Iraqi border. General Amos reported that "strike aircraft . . . and assault support/attack aircraft . . . are loaded with ammunition, fueled, and ready. Casevac [casualty evacuation] aircraft are forward . . . with the maneuver elements. Quick-strike package is identified. Crews . . . [are] on 30 min[utes] alert . . . [for] counter-fire mission should

DVIC DM-SD-04-09976

Manning their assault amphibian vehicle personnel (AAVP7A1), Marines from Headquarters Company, Regimental Combat Team 1, assemble as a convoy at a dispersal area prior to crossing into Iraq.

Iraqi forces attack . . . prematurely. . . . Commanders are briefed, targets assigned. We are prepared to execute the quick start option or the base plan; 3d MAW is ready to roll."[203]

Before long Conway received another order from CFLCC establishing D-Day and H-Hour for 1800Z on 19 March, when the air war would start. Ground operations were to start two days later, on 21 March.[204] Now everyone knew that war was hours away. The word was passed that there was a high probability of Iraqi missile strikes during the night of 19-20 March. Many Marines had their gas masks and chemical protective suits staged for a quick run to a shelter from their sleeping mats, but nothing happened during the night.[205]

The I Marine Expeditionary Force's war started the next day, 20 March, with an unexpected bang. At approximately 0725Z, there was a sound like that of a low-flying jet, followed by an explosion that shook the ground, and then by a tall gray-brown plume of smoke, about 200 meters north of the perimeter of Camp Commando. The very first effect was that it interrupted a staff meeting, General Conway and his principal staff members were in a briefing tent near the point of impact; they all dove under the tables.[206]* Other members of the I MEF staff wondered if the plume was poison gas, and if it was a terrorist attack, there had been no warning. Most reached for their protective gear, and NBC monitors swung into action. There was more than a little bit of confusion as Marines tried to figure out where to go and what to do. Many took shelter in the "Scud bunkers," the inverted concrete culverts and sandbag concoctions, but the combat operations center continued to operate. As many Marines sat jammed in the bunkers, with gas masks on, the word was passed to go to the highest NBC protective state (MOPP IV). Four Cobras were scrambled to scout for possible enemy attackers on the ground. By 0825Z it was clear what had happened and the "all-clear" had been sounded, but there were many more missile-raid alerts throughout the day, announced by the siren/loudspeaker combination known as "the Giant Voice." By

*I MEF Rear continued to function during the strike. As a result of the attack, I MEF Forward accelerated its preparations to move away from Camp Commando and, ultimately, into Iraq.

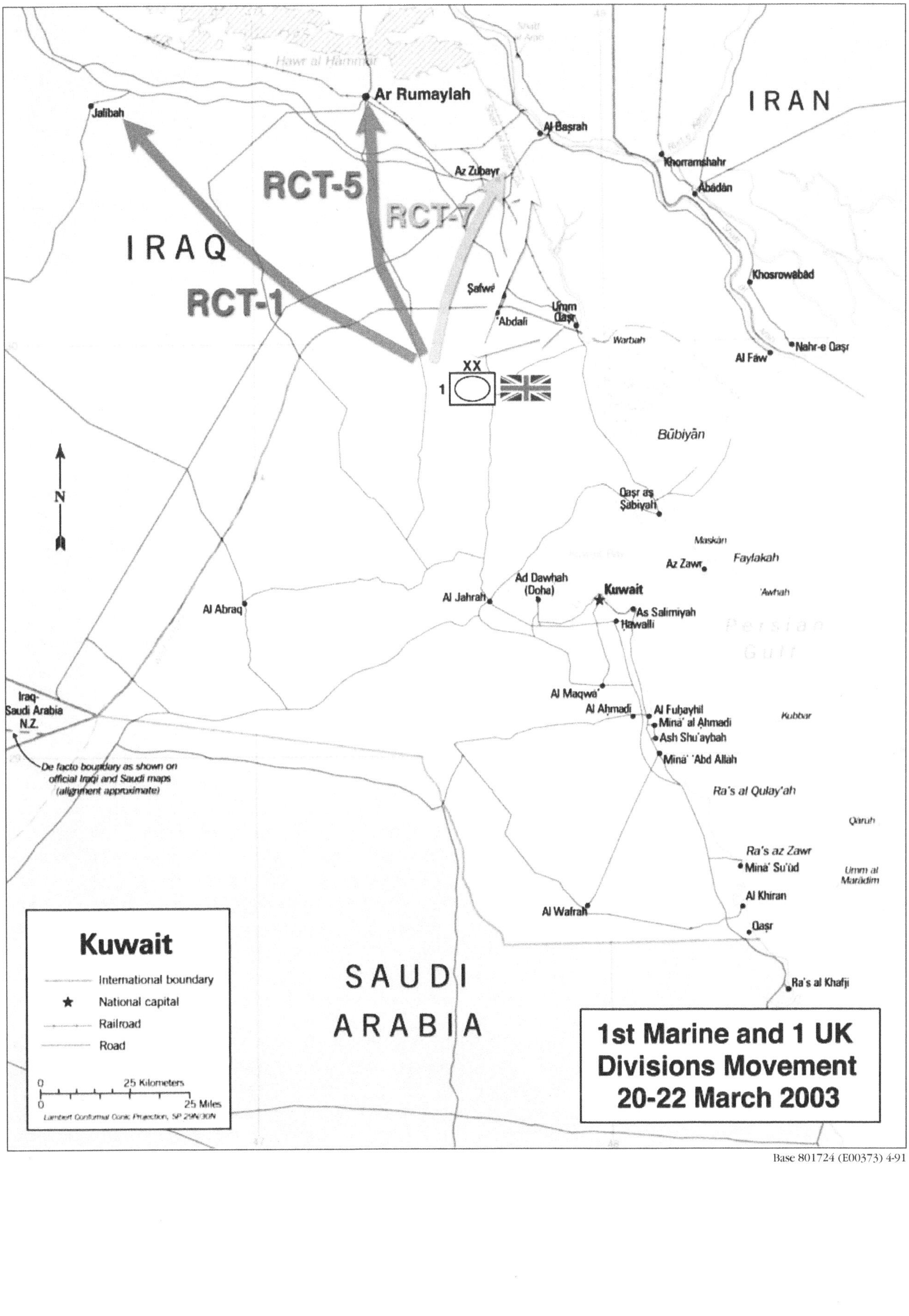

Base 801724 (E00373) 4-91

DVIC DM-SD-04-10906

Marine M1A1 Abrams main battle tanks with Company C, 1st Tank Battalion, line up and prepare to meet the enemy near Safwan Hill, Iraq, during the opening moves of Operation Iraqi Freedom.

one count, some 13 alerts had been sounded.[207]*

By now CFLCC had received its answer from CentCom and was free to attack into southern Iraq to secure the oil fields. Coalition Command passed the order to its subordinate commands to attack at 1800Z on 20 March, not on 21 March as CFLCC had ordered earlier.[208] (In General Conway's words, on account of "intel indicators that Iraqis had begun to destroy oil infrastructure, I MEF attacked into Iraq early.") It was around this time that General Mattis called Colonel Joseph F. Dunford, the commander of the lead element, and asked him how soon he could attack. Colonel Dunford asked for a few minutes to poll his staff, but soon came back with the answer—four hours. In fact, Regimental Combat Team 5 (RCT 5) was ready to go in three hours, and that is what they did.[209]

At 1512Z, I MEF released the execute order, and on 20 March at 1742Z, which was 2042 local, RCT 5's tanks crossed the border into Iraq in the dark, about nine hours ahead of the last regularly scheduled time. Instead of an attack at dawn, the regimental combat team was attacking at night, a much more complicated evolution, especially for a large, reinforced formation that was going into combat as a team for the first time. It was quite an achievement. When his troops crossed the border, General Mattis' official comment was "Tally-ho!"[210]*

It was only later that the Marines learned that during the night of 19-20 March, the United States had begun the war by hitting select targets in Baghdad, an unplanned bomb and missile strike at Saddam Hussein and his entourage, who, according to American intelligence, were spending the night in a bunker at a place called Dora Farms, a residential compound in south Baghdad. The report turned out to be false; a somewhat shaken Saddam Hussein soon appeared on Iraqi television vowing defiance. What was clear was that this was not an early start to the Coalition's long-planned air war. The timing of the missile attack on Camp Commando on 20 March suggests that the attack, and those that followed, was in retaliation for the Dora Farms attack.

What Dora Farms and CFLCC's images of burning oil wells did was breathe new life into the old debate about the separation between G-Day and A-Day. The CFLCC request for permission to attack early amounted to a request to reverse the order of G-Day and A-Day. When CentCom granted CFLCC's request, it was official: "rock and roll" for G-Day, wait one for A-Day. That left the relationship between air strikes

*The missile was most likely a Seersucker antiship missile, which literally flew under the air and missile defense radar, which is why there was no warning before the attack. It is not clear whether the warhead detonated; sources differ on this point. A Patriot antimissile battery brought down at least one other missile on 20 March. (I MEF Sitrep 191800Z to 201759ZMar03, copy in Reynolds Working Papers, MCHC, Quantico, VA).

*A related stimulus was an erroneous Central Intelligence Agency report that the Iraqis had moved a brigade of T-72 tanks into place near Safwan, just north of the border. This caused division to make an additional shift in its plans to accommodate the heightened threat of enemy armor. (I MEF sitrep 201800Z to 211759ZMar03, copy in Reynolds Working Papers, MCHC, Quantico, VA)

JCCC 030305-M-2237F-011

Veterans of combat missions over Afghanistan, two AV-8B Harriers from Marine Attack Squadron 542 taxi past each other on the runway at Ahmed Al Jaber Air Base, Kuwait, as they prepare for interdiction strikes against pre-planned targets in Iraq.

and ground attacks within I MEF's own sector. It is fair to ask whether there was to be any separation between them, and the answer was, not much at all, apparently for the same reasons. The Marines simply liked synchronicity, whether the context was the Iraqi theater as a whole or just the Marine battle space. As D-Day approached, the plan was for the 3d Marine Aircraft Wing, along with Marine artillery, to strike some targets in zone hours before the infantry went over the top. A variant of the plan called for even greater simultaneity, with a "spike" of close air attacks during the first day of the ground war. In either case, the guiding principle was coordination between wing and division, and everyone knew that.

Marines assigned to Battery I, 3d Battalion, 11th Marines, prepare to fire their M198 155mm howitzers against the Iraqi 51st Mechanized Division *and* III Corps Artillery *defending the Rumaylah oilfields.*

DVIC DM-SD-04-01580

As General Amos had written on 19 March, the "synchronization of major muscle movements is complete."[211]*

On 19 March, the 3d Marine Aircraft Wing had started attacking Safwan Hill, the high ground on the border between Iraq and Kuwait, which was a great observation point for the Iraqis, and was flying over the 5th Marines when they crossed the border into Iraq late on 20 March. On that day, the first full day of war, the wing flew 259 missions, 24 in support of CFACC, and 235 in support of I MEF, shifting its emphasis somewhat from generally "shaping" the battlefield to "preparing" specific objectives that division was about to assault. The arrangement with Coalition Forces Air Component Command was clearly working as the Marines intended. It is worth quoting the dry language of the wing's command chronology to get a sense of its activities on 20-21 March:

> Maintained constant airborne CAS [close air support] coverage in support of RCT 5 . . . Maintained constant F/W [fixed-wing] FAC(A) [forward air controller (Airborne)] coverage for both RCT 5 and RCT 7, to enable interdiction of enemy counterattack or reinforcing elements. Conducted F/W counter fire in support of RCT 7 . . . Began . . . effort in earnest against Iraqi 2d echelon forces, focusing on enemy indirect fire and SSM [missile] assets. . . . Shaping MEF battle space . . . Focus on MRL [multiple rocket launcher], artillery, and reinforcing armor in Al Amarah and Basrah areas . . . Provided 8 F/W sorties to conduct CAS in support of UK forces engaged in Al Faw . . . Pushed . . . units forward to commence establishment of FARPs at Safwan and Jalibah.[212]

The last item does not look particularly dramatic on paper, but the words belie an impressive accomplishment. Ultimately, some 4,000 ground personnel from the wing crossed the berm and, according to plan, set up some 15 small air bases and support points in Iraq. The concept was not new, but the scale was, as was the speed and flexibility of execution.[213] There were some dramatic, even heroic, moments for the support Marines who made it all happen. One of the most memorable interviews any field historian conducted during the war was that of Gunnery Sergeant Melba L. Garza, the operations

Photo courtesy of 15th MEU

With surrendered Iraqi soldiers near the port of Umm Qasr at a safe distance, a machine gunner with Battalion Landing Team 2d Battalion, 1st Marines, carefully guards the prisoners before turning them over to special handling teams.

chief of Marine Wing Support Squadron 271. She recounted, in a deadpan voice, her memories of traveling north in one of the long convoys of support vehicles, which included a number of fuel tankers. These slow-moving convoys stretched literally for miles along the few highways through the desert. When the convoy was ambushed by Iraqis on the ground, there was not a great deal that any one individual, especially someone like an operations chief armed with a 9mm pistol, could do about it except hope that the accompanying Cobras would be able to deal with the enemy, that is, until the Cobras ran low on fuel and the support Marines decided to do some "hot" refueling on the spot so that the Cobras could stay in this particular fight until it was over. Refueling is normally done in a controlled environment, after the aircraft powers down. If time matters, it is all right to refuel "hot," while the blades are still turning. But there is no practice for refueling under fire. To say the least, this group of Marines redefined the term "hot."[214]

Within about 11 hours of crossing the line of departure, RCT 5 had seized most of its initial objectives. This was largely because the division was so well prepared and coordinated on many levels. A look at fire support coordination from an artilleryman's perspective suggests only some of the complexities:

> During . . . the "Opening Gambit," the opportunity for fire support coordination to break down was at its greatest. Consider managing a fire support coordination line shift, a battlefield coordination line . . . shift, coordinated fire lines

*1st Marine Division planners considered the air war and shaping in the I MEF area of operations to be two separate issues. For the most part, the targets in the "shock and awe" air war were far from that area of operations.

> . . . shifting up to seven times, opening and closing multiple keypad variations of up to six different killboxes [the map grids used to coordinate fires], coordinating numerous no-fire areas . . . and managing a restricted target list of over 12,000 targets all within a matter of 12 hours. [At the same time there was coordination] . . . with a counterobservation post program of fire, breaching operations, three regimental combat teams attacking at separate times. . . . two counterbattery programs of fire, one counterarmor program of fire, attacking high-payoff targets of opportunity . . . a transfer of control between the division main . . . and division forward . . . and deteriorating weather conditions.[215]*

On the same day, 21 March, the rest of the Marine division, as well as the British division, poured through the breaches in the border installations between Kuwait and Iraq on their way north, while the "official" air war, the "shock and awe" phase, finally began over Baghdad with the obliteration of a carefully chosen set of targets. The world watched on live television. This was certainly an impressive display of precision targeting and pyrotechnics that lit the night sky and offered one very fine photograph opportunity for the news services. Some of the resulting photographs of Baghdad in flames will symbolize the war for years to come. But it did not spark an uprising, and the regime did not collapse.[216]

Back in the south, there was sporadic fighting, some of it sharp and deadly, which continued into 22 March, when RCT 7 engaged isolated pockets of resistance in the Marine area of operations. One of the first Marines to be killed in Iraq died in one of these firefights. He was Second Lieutenant Therrel S. Childers of 1st Battalion, 5th Marines, who was hit when Iraqi fighters in a civilian pickup truck attacked his unit in the oil fields. But by and large, the Marines were pleasantly surprised by the relatively light resistance they encountered, which was much less than they had expected or feared. The local Iraqi command, the *51st Mechanized Division*, had ceased to exist in any recognizable military form, and there were already hundreds of apparent deserters in civilian clothes fleeing to the west on 21 March. While some of the Iraqi wells near the border with Kuwait burned brightly after being sabotaged by their owners, there were also instances of Iraqi oil company employees waiting patiently for the Coalition forces to arrive, after carefully following the instructions for preventing sabotage in the CFLCC leaflets that rained down from the sky.* Overall, there was far less destruction to the oil infrastructure than many had feared. That was the good news. The bad news was that the oil infrastructure had not been maintained for years and was in terrible shape. By the end of the day, with most of I MEF's initial objectives having been seized, the division was getting ready to turn them over to the British. The official relief in place occurred without incident on 22-23 March. When reporting the relief, division added the note that after receiving information about possible Iraqi infiltrators in American uniforms, General Mattis had "directed all division Marines to remove their moustaches as part of the . . . effort to distinguish Iraqi infiltrators."[217]

With the British now protecting its right flank, the division could now turn west, moving in the same direction as Task Force Tarawa, which had rolled through the breach and crossed into Iraq on the early morning of 21 March and moved toward the town of Jalibah, paying special attention to Jalibah Air Base, where I MEF was soon to place its forward headquarters, and the key terrain at the small city of An Nasiriyah.

*In this war, many artillery missions, like counter-battery fire, were largely computerized and completed in a matter of seconds.

*This was later touted at a Coalition Forces Land Component Command briefing attended by the author in late March.

Chapter 5

The Bridges of An Nasiriyah

The heart of An Nasiriyah is something like an island between two waterways, the Euphrates River on one side, running roughly northwest to southeast, and the Saddam Canal, which runs more or less parallel to the Euphrates on the other, eastern side of the city. An Nasiriyah controls the bridges over both river and canal that lead to Route 7, the main highway to the north through the center of southern Iraq. Route 1, a more westerly highway to Baghdad, passes within a few miles of An Nasiriyah but does not come within the city limits. Apart from the fact that if you approach from the desert, An Nasiriyah is like an oasis with its palm trees and other greenery, but the city has little to offer; pictures show an uninviting, Third World "sprawl of slums and industrial compounds," with two to three-story concrete buildings set on a grid of bad roads and alleyways, many strewn with garbage and raw sewage.[218] The city was all the more rundown because its largely Shia population was known to have opposed Saddam's rule, and he repaid the favor by neglecting even its most basic needs.

Task Force Tarawa's mission was to be the first Marine unit at An Nasiriyah and to secure the bridge over the Euphrates on Route 1, which lay a few kilometers west of the city. It had the follow-on mission to "be prepared to" secure the bridges over the Euphrates and the Saddam Canal on the eastern edge of the city, which the Marine division, especially Regimental Combat Team 1 (RCT 1), would need to pass over on its way north on Route 7 toward Al Kut. The plan was for RCT 1 to more or less keep pace with RCTs 5 and 7 as they moved up Route 1. In a sense it was a straightforward mission, and it was one that had been assigned to Task Force Tarawa before it arrived in Kuwait. Task force officers, down to the company level, had performed map studies, war games, and rehearsals for An Nasiriyah.[219] That said there was a difference between the East and West Coast Marines, who had been visualizing the first days of the war since the summer of 2002. By contrast, most Task Force Tarawa Marines did not begin to focus in on the mission until December 2002, and while the operational picture had come into progressively sharper focus as Ground Day approached, there were still a lot of unknowns about An Nasiriyah. One lingering question was just how "permissive" the city would be. The general assumption was that the Marines would receive a friendly welcome from the townspeople, and at worst some light resistance from Saddam loyalists. Certainly the U.S. Army, which would pass through part of the area before the Marines, would find out for sure.[220]

On 22 March I MEF announced that within the next 24 hours it wanted to secure the eastern crossing sites at An Nasiriyah and commence the forward passage of lines by 1st Marine Division through Task Force Tarawa.[221] On 23 March, I MEF reported that it had released its Fragmentary Order 017-03 that tasked Task Force Tarawa with conducting a relief in place with the 3d Infantry Division "at [the] Highway 1 Euphrates River crossing and attack to seize [the] bridges east of An Nasiriyah ... [in order to] facilitate the unimpeded continuation of the attack by 1st MarDiv to the north and northwest."[222]

It was a mission fraught with potential complications. Task Force Tarawa's western boundary was with the Army, and the task force's first job was to relieve elements of the 3d Infantry Division that had passed through the area on their way to the western desert. Then, after seizing the bridges that passed through the city, the task force would have to coordinate the forward passage of lines with RCT 1. These missions are difficult enough in peacetime between units that have trained together, let alone units from two separate Services, Army and Marines, and two separate chains-of-command, task force and division, which literally had trouble getting on the same radio frequency. There was a final complication: Marines, many of them new to combat, not to mention the demands of combat in a city, were beginning to get very tired. The burst of adrenaline that had carried them across the border and into combat could not keep all of the Marines on their feet forever, especially when everyone, from the force commander to the private on the firing line, had to work and fight in the hot, bulky NBC protective gear, usually with a flak vest as the outer layer, in the desert heat. The Marines were approaching a culminating point of sorts, the end of the first phase of the battle for Iraq.[223]

At An Nasiriyah, Task Force Tarawa was, briefly,

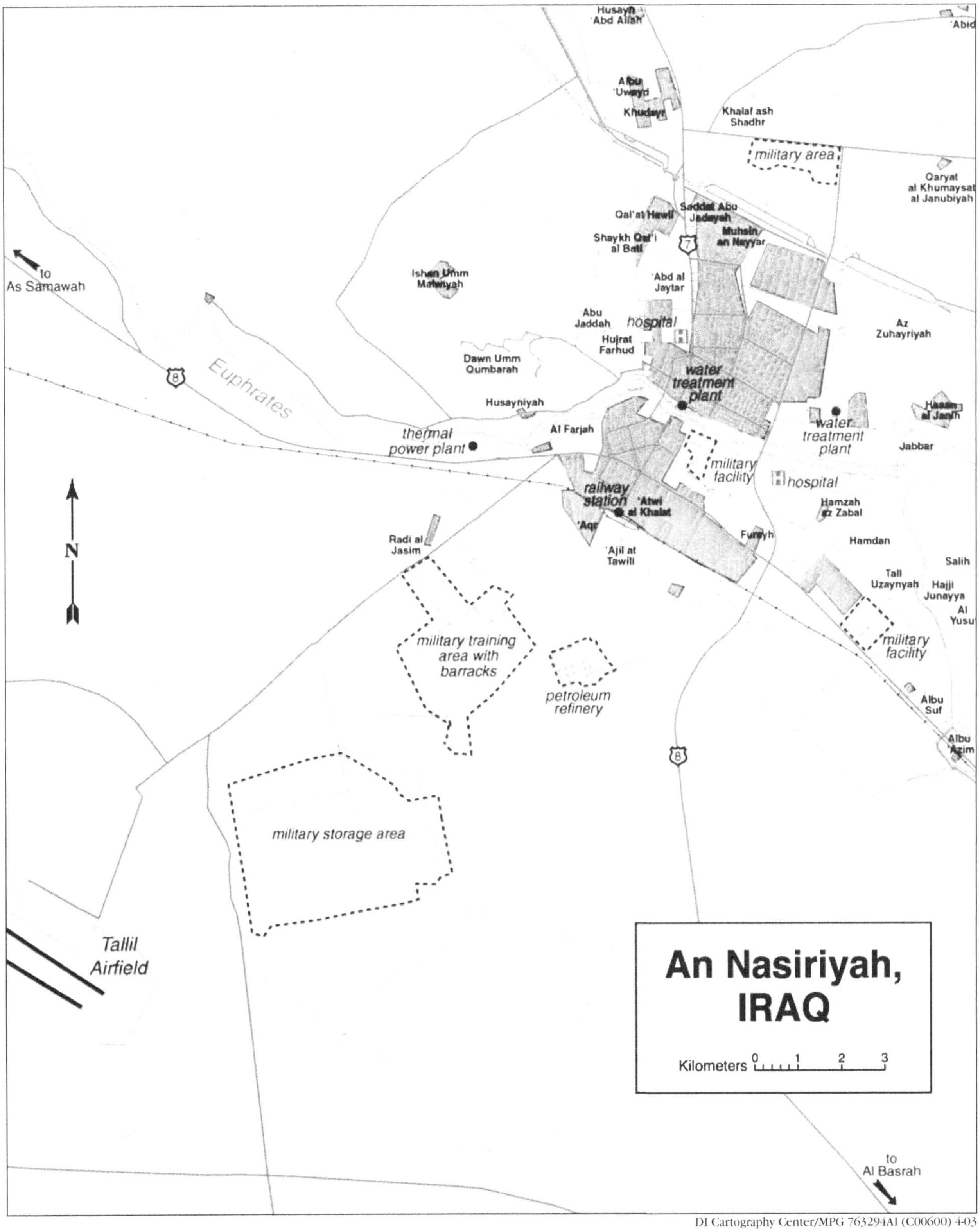

DI Cartography Center/MPG 763294AI (C00600) 4-03

the de facto main effort, and the burden of winning this particular fight fell on the shoulders of Brigadier General Richard F. Natonski. He has been described as a large man with a deliberate, confident bearing who was shaped by his experiences as an expeditionary unit commander and as a senior staff officer in the current operations section at CentCom as well as in Plans, Policies, and Operations at Headquarters Marine Corps. Senior members of Natonski's staff liked working for him and even called him a "model

DVIC DM-SD-05-04640

BGen Richard F. Natonski, here speaking to a member of the media, urged the Marines of Task Force Tarawa to press on and quickly seize and hold the bridges in An Nasiriyah, to keep the enemy off balance.

commanding general." An Nasiriyah was to be Task Force Tarawa's first major challenge. Both Natonski and many of his Marines knew it might be their only time at center stage. He was determined to get it right, even if that meant demanding sacrifices from his line commanders.

When they met late on 22 March, General Natonski and the commander of RCT 2, Colonel Ronald L. Bailey, focused on the I MEF order. Following up on the I MEF fragmentary order, General Natonski tasked Colonel Bailey, who commanded most of the ground troops in the task force, with seizing both the Route 1 and the Route 7 crossings on 23 March. This was a departure from the original plan to begin by seizing the Route 1 bridge to the west of the city, and then move on the "eastern," or "city" bridges "on order," which meant there would be a delay between the two evolutions, a chance, however short, to fine-tune the planning for the next step. There was nothing new by way of intelligence about An Nasiriyah, but Colonel Bailey was warned to expect small arms fire. When he expressed some concerns, his troops needed rest and his mechanized assets were low on fuel, General Natonski told him it was important to press on, the Marines would have to run on adrenaline.[224]

Marines throughout the theater felt that the next day, 23 March, started bad and never got any better. Around daybreak, the U.S. Army's truck-borne 507th Maintenance Company lost its way and blundered into a bloody ambush in An Nasiriyah. One of the members of the 507th was Private First Class Jessica Lynch, who was wounded and captured by the Iraqis and after being rescued went on to become a celebrity of sorts. This helped set the stage for the events that followed, as did the relief in place to the west of the city between Task Force Tarawa, in particular Company C, 2d Light Armored Reconnaissance Battalion, and the 3d Infantry Division. This was accompanied by a boundary shift that put virtually all of An Nasiriyah in Marine battle space. The relief in place and the boundary shift were otherwise unremarkable, despite occasional allegations to the contrary.[225*]

A few miles away, during the advance toward An Nasiriyah from the southeast, at about the same time as the 507th was ambushed, 1st Battalion, 2d Marines, with tanks and a combined antiarmor team in the lead, encountered small arms and mortar fire while still well outside the city, a portent of more to come. A short while later the Marines encountered a few survivors from the 507th, which fed the hope that there might be more survivors up ahead. This possibility put added pressure on Colonel Bailey.

Mid-morning on 23 March, General Natonski flew twice to the battlefield in his command Huey helicopter. When he looked down from the air, he did not see the regimental combat team's troops where he wanted them to be. His impression was that the attack was not going quickly enough; what were they waiting for? He ordered the Huey to land at Colonel Bailey's forward command post, which was near a railroad bridge outside the city limits, so that he could urge him to move faster. On the first visit, the general spoke both with Captain Troy K. King, USA, the commander of the 507th, who told him firsthand about the ambush, and with Lieutenant Colonel Ricky L. Grabowski, the commander of the lead battalion, 1st Battalion, 2d Marines, in addition to Colonel Bailey himself.[226] General Natonski told Colonel Bailey and Lieutenant Colonel Grabowski that he could see nothing in their way, no enemy tanks or other mechanized assets, and that the force was relying on Task Force Tarawa to seize the bridges and hold them open. General Natonski believed that by moving fast, the task force would keep the enemy off balance and, ultimately, limit the number of friendly casualties. After talking to Colonel Bailey, General Naton-

---

*Reacting to a question about boundaries, LtGen David D. McKiernan commented that the events in An Nasiriyah did "not equate to any seam or any joint problem. There were on-order boundaries that were placed in effect both south and north of An Nasiriyah, between V Corps and I MEF, which made sense [and] which were triggered at the right time . . . I don't think the boundary shift could have gone much better." (LtGen David D. McKiernan intvw, 30Jun03 [U.S. Army Center of Military History, Washington, D.C.])

DVIC DM-SD-05-04644

The bridge spanning the Euphrates River at An Nasiriyah needed to be secured intact as it was on the vital main supply route for Coalition forces moving north in Iraq.

ski flew to the I MEF command post at Jalibah, a few minutes' flying time from An Nasiriyah, and briefed General Conway on the situation. When he returned to Bailey's position an hour later, Natonski repeated his orders, which now contained the even more forceful pronouncement, seemingly from Conway's mouth, that Bailey was holding up the force.*

Colonel Bailey did as he was told and made the general's intent his own, despite personal reservations. This was, after all, a city that had not been thoroughly probed by reconnaissance in the recent past, perhaps because of the expectations that the Iraqis would be friendly.** Similarly, there appears to have been no plan to conduct preparatory artillery or air attacks before the Marines entered the city limits, even though air and artillery support were on call and became very active participants in the battle. Cobras flew above the advancing Marines, and the artillerymen of 1st Battalion, 10th Marines, were following in trace of RCT 2's lead elements, ready to emplace and process fire missions in very short order. General Natonski later praised the battalion for providing "invaluable" counter-battery fire during the battle, in addition to responding for calls for fire from Marines under attack. [227]

After the first meeting ended and the three principals went their separate ways, Colonel Bailey had second thoughts and wanted to talk to Lieutenant Colonel Grabowski again. He wanted to be sure that Grabowski had not left with the wrong impression–Grabowski should proceed, but not at all costs. It was the kind of thought that passes through a commander's mind as he sends his troops off to battle, especially on the first day. Bailey was a conscientious officer with a reputation for taking care of his Marines. But Grabowski was already out of reach, moving into the attack, and it was too late to review the bidding. [228]

It was now sometime before noon; after the fight, Lieutenant Colonel Grabowski commented that it was difficult to remember exactly what happened when, given the intensity of the combat, which plays tricks on a person's sense of time. The sun was high, the day clear and hot. Two kilometers south of the Euphrates, the tanks detached to refuel, but the rest of

---

*An additional consideration BGen Richard Natonski mentioned in a postwar interview was that the cluster of American forces south of An Nasiriyah made an excellent target for Iraqi weapons of mass destruction, a threat that was still very much on everyone's mind at the time. It is not clear whether he mentioned that point to Col Bailey on 23 March. (BGen Richard F. Natonski intvw, 26Mar04 [MCHC, Quantico, VA])

**To be sure, Marine reconnaissance elements close to the western bridge were working their way back toward Task Force Tarawa, and U.S. Army Special Forces were operating in the area on each side of the river, but this did not equate to a reconnaissance of the city itself. (Reynolds, Journal, entry for 1Jul04, recording a conversation with Task Force Tarawa's G-3; BGen Richard F. Natonski intvw, 26Mar04 [MCHC, Quantico, VA])

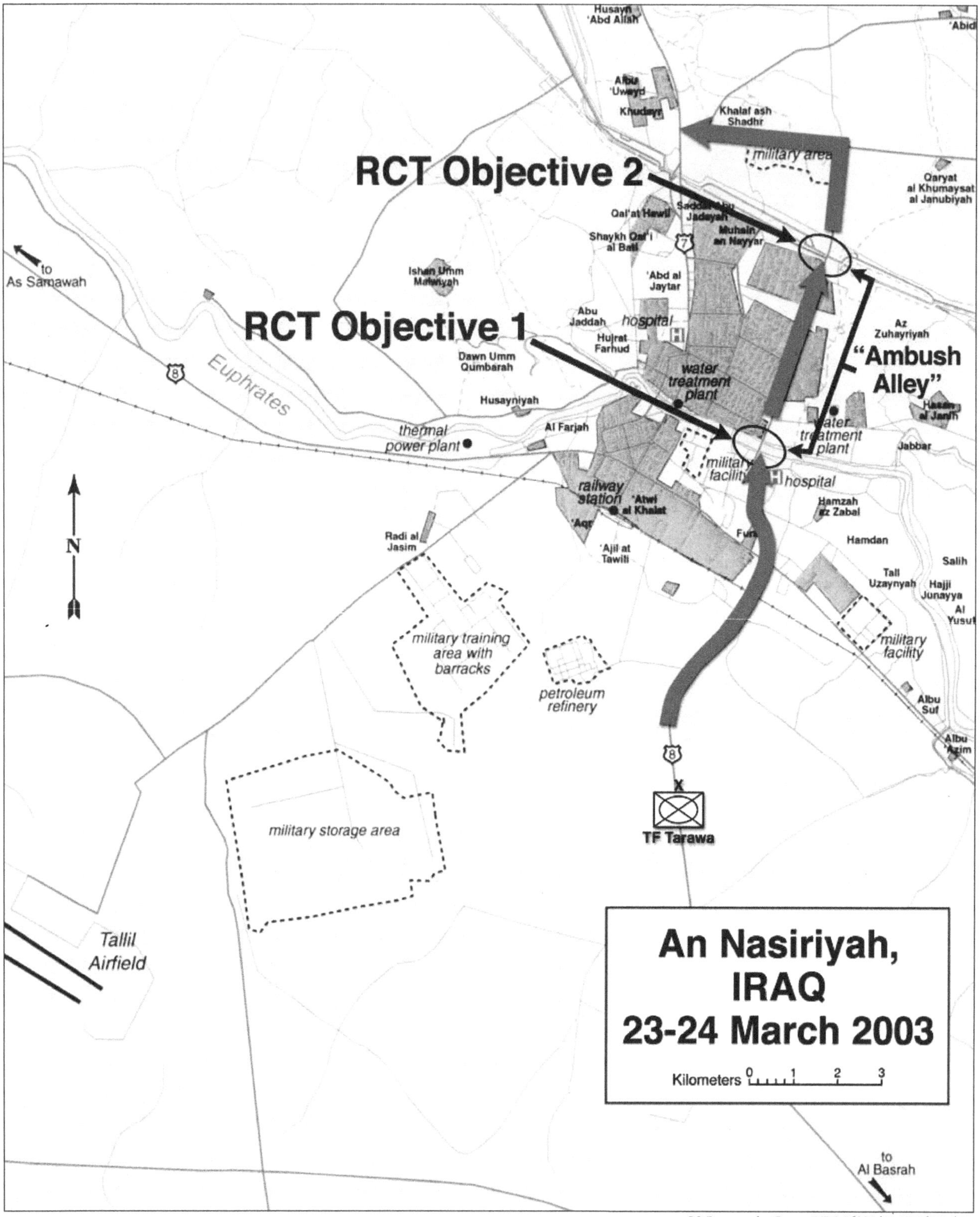

DI Cartography Center/MPG 763294AI (C00600) 4-03

Grabowski's battalion continued over the river and into the city. After the friendly tanks detached to refuel, Company B encountered enemy tanks, which were engaged and destroyed by antiarmor Marines, including a "Javelin" team. Company A secured the far end of the eastern bridge over the Euphrates. Then Company B, with the battalion command group, crossed over the bridge and, looking for a route around downtown An Nasiriyah, drove onto apparently firm ground that turned out to be a kind

DVIC DM-SD-05-04643

Objective 2 for Task Force Tarawa was the seizure of the bridge crossing the Saddam Canal. The road between the two bridges in An Nasiriyah became known as "Ambush Alley," because of the intense enemy fire experienced by the Marines as they traversed the four kilometers of cityscape.

of tarry quicksand. One Marine, Corporal Jason J. Polanco, was looking at the two tanks in front of him that were running level with his own vehicle one second, and then, in the next, "just dropped in[to] the mud."[229]* Company B's plan to come up on the southeastern flank of the bridge over the Saddam Canal, and to support the assault on that bridge by fire, was on hold.

With Company B stuck on its right to the east, Company C forged ahead through four kilometers of cityscape that became known as "Ambush Alley," coming under "intense machine gun, small arms, and RPG fire" from a variety of combatants—a mix of regular soldiers and paramilitary fighters—almost all of whom wore civilian clothes. Sometime after noon, mindful of the pressure on the task force and the regimental combat team, Company C's commander, Captain Daniel J. Wittnam, decided to keep moving ahead. He appears to have made this particular decision more or less on his own, although it was definitely consistent with his battalion commander's intent. Lieutenant Colonel Grabowski had effectively conveyed his determination to seize and hold the bridges. Not only was Wittnam's the first Marine company over the Saddam Canal, but it was now out ahead of the rest of the battalion, more exposed to the enemy than anyone else.[230]

Wittnam's company drove across the wide, flat modern bridge; it looked as much like a stretch of

A humvee with a 7.62mm medium machine gun races to aid fellow Marines engaged in a firefight in An Nasiriyah. Mounted combined weapons teams were formed from the heavy weapons company of the infantry battalions.

JCCC 030401-M-5977R-010

*This may have been Sobka, a geological phenomenon peculiar to the Middle East, which was encountered elsewhere in Iraq and mired other Marine vehicles. It may also just have been sewage, with the top layer baked into a crust.

highway as a bridge, into what amounted to a kind of fire sack. The weapons platoon sergeant, Staff Sergeant Lonnie O. Parker, remembered the feeling: "We all came out of the door [of our amtrack], got . . . situational awareness of where we were, where the enemy was. They were located north of us, they were located to the west of us, they were located to the south of us, and they were located to the east of

The Other Ambush on 23 March

A few hours after Wittnam's Marines drove across the bridge over the canal into the enemy fire sack, another group of Marines had a similar experience in the desert north of the city. The story begins late on the afternoon of 23 March. One of division's Lockheed P-3 Orions, borrowed from the U.S. Navy, was flying high and slow above the battlefield to scout the route up Highway 1, while light armored reconnaissance Marines did the same on the ground. Although one of the accompanying Cobra pilots who had been conducting low-level scouting for the armored reconnaissance Marines thought he had seen some signs of enemy activity, the route seemed clear to division. General Mattis decided he wanted the Marines to move even faster, and personally called one of his armored reconnaissance commanders, Lieutenant Colonel Herman S. Clardy III, of 3d Battalion, codename "Wolfpack," to tell him to proceed up to the town of Hantush some distance up the road. Clardy told his commanders to pull off the road and gather around him for a five-minute briefing and an order. Minutes later, the battalion went "screaming north" into the gathering dusk, well out ahead of any friendly formations and outside the artillery fan.

Captain Charles J. Blume, Clardy's fire support coordinator, remembered in vivid terms what happened next:

> We definitely could feel [that] we were getting well out in front of the division. We lost communications with the DASC-A [the airborne direct air support coordinator] and it was starting to get dark. . . . We began to see abandoned weapons and equipment strewn along the highway. [We saw a] suspicious vehicle . . . to our front that [looked at] . . . us and sped away. . . . We could all feel the hair standing up on the backs of our necks. You could tell something was about to happen.

What did happen next, at 1607Z, was burned into the memory of one of the company commanders, Major Bruce Bell:

> The Fedayeen had actually laid out a decent "U"-shaped ambush spread over . . . 500 meters on both sides of the road. . . . They picked a tactically sound, defensively oriented bend in the highway . . . to exploit massed surprise . . . fires on the lead units of whoever fell into the trap. They also had assembled a column of approximately 10 tanks, armored personnel carriers, and other various "technical" . . . vehicles [mostly pickup trucks with machine gun mounts] which they positioned on the eastern flank of the ambush position, hoping to use a north-south jeep trail . . . to move down and flank units caught in the kill zone on the highway.

When the enemy opened fire, at first it was only scattered tracers flying across the road, and then there was a torrent of fire all up and down the route of march. Some of the enemy appeared to be massing for an attack. The Marines fired back, but the enemy fire kept coming. The air officer tried to reach Clardy over the radio, but could not. Someone, either the air officer or the communications officer, called "Slingshot," the heart-stopping code word in this war for "I am being overrun." The division later reported that the call came over the Iridium cell phone, the official/unofficial alternate communications system in this war, and that "3d MAW immediately responded with 6 Harriers and 4 Cobras, followed shortly thereafter by a host of additional air assets." It was all over by 1741Z, the remnants of a battalion-sized Iraqi unit left smoking on the battlefield. The Marines, miraculously, emerged from the fight with one wounded in action and some battle scars on their vehicles, but were still able to continue moving up the road.*

*IMEF sitrep 221800ZMar03 to 231759ZMar03 (Copy in Reynolds Working Papers, MCHC, Quantico, VA); 1stMarDiv ComdC, Jan-Jun03 (GRC, Quantico, VA), sec 2, chap 5, pp. 5-8; Maj Bruce Bell, e-mails to author, 11-12Jul03 (Copies among Reynolds Working Papers, MCHC, Quantico, VA)

DVIC DM-SD-05-04642

A Marine surveys some of the damage done to the city of An Nasiriyah after more than eight days of fighting. The heavy cost of the fighting impacted the lives of both the populace and combatants for some time to come.

us."[16] The Marines found themselves on a roadway on the far side of the canal, surrounded by fields that were lower than the roadway, which made them good targets for the waiting Iraqi soldiers. They could see the enemy scurrying in and around their fighting positions and the plain concrete buildings that were all some distance away on the other side of the fields and the berms that helped to define the fields. Apparently emboldened by their success against the 507th, and now ready to fight the next wave of Americans, the Iraqis fired infantry weapons and, especially, mortars up to 120mm in caliber. Marines later learned that the enemy's positions were primarily oriented to the north, in order to defend against an airborne attack that was never planned, let alone executed. The enemy infantry showed no interest in closing with the Marines and getting within easy range of their small arms; they chose either to fire from their foxholes or to dart out from a courtyard or alleyway to fire off a few rounds. At least one Marine small unit leader had to keep his hard-chargers from rushing across the open fields at the enemy.[232]* Instead, the Marines used their own weapons against the enemy and called in artillery; 1st Battalion, 10th Marines' guns soon fired to good effect. Sadly, the company's forward observer, First Lieutenant Frederick E. Pokorney, Jr., was killed while calling for fire. But thanks in part to the enemy's poor marksmanship, and thanks in part to the Marines' good work, the company was soon making some headway against the enemy and consolidating its own position.

Captain Wittnam had seized an important objective, and he wanted to hold it until reinforced or relieved. Then, a U.S. Air Force Fairchild-Republic A-10 Warthog swooped by, circled, and lined up for a strafing run on the Marines as they watched in horror. Although a jet, the Warthog is designed to fly low for close air support missions and, with its depleted uranium rounds, was known as a good tank buster. It is usually a welcome sight on the battlefield and had already done good work on 23 March against other targets. But now it was bearing down on friendlies.

One Company C Marine on the bridge, First Lieutenant Michael S. Seely, had been strafed by an A-10 before, in 1991, and he knew instantly what was happening:

> I did not even have to look up, because I knew exactly what that sound was. . . . I ran up and found 2d Platoon scattered all around the area there, but I grabbed their [radioman and] said, "Put that damn thing on battalion Tac now!" I got on battalion Tac immediately and started calling, "Cease fire! Cease fire!" Timberwolf 6 [the battalion commander] came up, perfect[ly] calm, and I started talking to him. He said, "What do you got?" I said, "We [are] having friendly air, [an] A-10 strafing our pos." I do not know the time that it took, but it was probably a couple minutes later . . . I do not know, 10, 15, whatever–the A-10 was still circling overhead.[233]

To make matters worse, the A-10, apparently along with his wingman as the A-10 was flying in a two-plane section, made numerous deadly passes, while the Marines on the ground tried every way possible to end what arguably became the most notorious friendly fire incident of the war. By mid-afternoon each of the battalion's rifle companies was, in the understated words of the *Marine Corps Gazette* article about the fight, "decisively engaged in non-mutually supporting positions" throughout the

*SSgt Lonnie O. Parker remembered that the rocket propelled grenade and small arms fire soon tapered off. (SSgt Lonnie O. Parker intvw, 23Mar03 [MCHC, Quantico, VA])

JCCC 030411-M-6910K-110

Marines assigned to Combat Service Support Battalion 18 work to retrieve a destroyed P7 amphibious assault vehicle following the fighting in An Nasiriyah.

city and only in sporadic touch with their battalion commander, whose tactical radio nets have been described as "clogged."[234] It was only later in the afternoon that the companies were able to support each other, which is what Wittnam had been waiting for, given his intent to hold the position until reinforcements could get to him. He estimated later that the wait was between two and a half and three hours.[235] He and his men had breathed an enormous collective sigh of relief when a pair of Marine tanks rumbled into their lines and suppressed the remaining Iraqi opposition once and for all. There had been a couple of anxious moments when the Company C Marines first heard tanks coming their way, and before they identified them as the friendly reinforcements they were hoping and waiting for.[236] Exhausted, dirty, and bloody, they began to recover from the first day of heavy fighting for I MEF in Iraq, which cost the lives of 18 Marines. With some pride and some sadness, Staff Sergeant Parker summed it all up when he said it "was not supposed to be no really big conflict that day," but "we put up one hellacious of a fight. . . . [I]t is really sad when it ends and you lose the majority of your people not from enemy fire but from friendly fire."[237]

Task Force Tarawa remained heavily engaged in An Nasiriyah for eight more days, working to clear the enemy from the route through the city. This included some bitter house-to-house fighting, defeating some one thousand enemy fighters massing for a counterattack and, on 1 April, support for the mission to rescue that most famous of survivors of the ambush of the 507th, Private First Class Lynch. The field historian attached to Task Force Tarawa commented that 3d Battalion, 2d Marines, became "adept at collecting front-line intelligence and following up with what were termed 'House Calls' on the homes of officials of the regime," which, in turn, led to further contacts and a growing hold on An Nasiriyah.[238] Throughout this period there was good cooperation with U.S. Army Special Forces, whose detachments continued to operate alongside the task force and produced actionable intelligence, which often led to fire missions, air strikes, or raids like the Lynch rescue.[239]* It is worth noting one such attack, early on, when Task Force Tarawa drove a band of *Fedayeen* from a hospital that turned out to have a stockpile of "200 AK-47 . . . [rifles], 20,000 rounds of ammunition, 3,000 chemsuits and masks, a tank [!]. . . , 400 Iraqi uniforms, and four U.S. Army uniforms."[240] This was one of the first concrete indications to the Coalition that Iraq was one vast ammunition dump-cum-armory, which would pose a disposal problem on an unimagined scale. Nevertheless, by early April, Nasiriyah was taking its first tentative steps, under Task Force Tarawa, toward post combat reconstruction, a few days ahead of the rest of Iraq.

Various controversies about 23 March were to continue for some time. There were questions, some of which made more sense than others, about such things as why the Marines did not bypass An Nasiriyah or in general do a better job of fighting in cities.[241] There were other questions about whether it was right for Task Force Tarawa to push into An Nasiriyah the way it did. Would it have made sense to wait for the tanks moving with 1st Battalion, 2d Marines, to refuel before advancing into the city? Or to wait until the battlefield had been "shaped," that is, why had not there been a more thorough reconnaissance before the Marines entered the city, or why had not artillery and air struck Iraqi positions before the Marines reached them? Did Captain Wittnam's decision to advance Company C by itself make sense? There are answers, some more compelling than others, to all of the questions about 23 March. General Natonski's pressure on RCT 2, and Captain Wittnam's decision to push on and then hold on, may have saved the day. One of the Iraqi commanders captured at An Nasiriyah commented that the fast tempo of the American advance had made it impossible for him to respond in time and that he had been "shocked" at the aggressiveness of Marine small unit

*This was representative of cooperation with various Special Forces in the entire Marine area of operations, although Gen Mattis and Gen Conway eventually came to the conclusion that these operators were better at some tasks than others. In particular, they complained it could take them too long to plan a mission to take advantage of a rapidly breaking situation.

DVIC DM-SD-04-01582

A Marine comes to the aid of injured and displaced Iraqi civilians caught in a firefight north of An Nasiriyah. The civilians were later evacuated to the triage area of Regimental Combat Team 1 to receive medical treatment.

leaders. "He said that his fighters were very confident initially . . . but became dispirited when the Marines kept coming at them."[242*]

From the CFLCC and I MEF point of view, it would not have made sense to bypass An Nasiriyah. The layout of the roads and bridges around the city made it difficult to bypass. Not even assault bridging would have helped. When RCT 1's Colonel Joseph D. Dowdy considered that option, he concluded it would add days to the journey north. Even if the bypass option had made sense, the Marines would still have had to find a way to deal with potential enemy threats to their lines of communication from within the city. Simply put, CFLCC forces needed to control all of the routes in and around An Nasiriyah and had had little choice but to go through the city. As General McKiernan put it: "Everybody had to go by An Nasiriyah, in either corps' sector, because that was the only place to cross the Euphrates. . . . It was just the [nature of] the whole fight in the south. Our enemy concentrated out of urban areas."[243]

Perhaps the most bitter controversy about An Nasiriyah was in a class by itself, the controversy over the "friendly fire" by the A-10s, which led to a lengthy investigation and, in April 2004, to the release by Central Command of a 900-page report that concluded it was a Marine air officer who had cleared the A-10s to fire on any vehicles on the far side of the

*There was supporting air on station and artillery on call throughout the period; the issue is whether there was an adequate amount of preparatory fire.

DVIC DM-SD-04-01589

A convoy of humvees from Weapons Company, 1st Battalion, 5th Marines, with tube-launched, optically-tracked, wire-guided (TOW) missile launchers, makes its way north through the "mother of all sandstorms."

bridge over the canal. Located a few hundred yards away with Company B, the officer apparently believed he was with the lead element of the battalion and therefore no Marines were in front of him and Company C did not have its own forward air controller. The report essentially invited the Marine Corps to continue the investigation to determine whether any disciplinary action should be taken against that officer, which would perhaps prolong the controversy. One observer has argued that it is unfair to single out the air controller, since his actions followed, at least in part, from the actions of others in his chain-of-command, and since it appears that some of the battalion's communications nets failed at crucial points.[244] Many Marines found that line-of-sight communications inside the city limits was terrible.

The end result of the general confusion on 23 March and the attack by the A-10 was not only the painful casualty count but also a monumental traffic jam that lasted through 24 March and into 25 March.[245] The CentCom report stated that "eight of the deaths were verified as the result of enemy fire; of the remaining 10 Marines killed, investigators were unable to determine the cause of death as the Marines were also engaged in heavy fighting with the enemy at the time of the incident. Of the 17 wounded, only one was conclusively . . . hit by friendly fire."[246] Behind the Marines in the fight at the bridge and along "Ambush Alley" 30 kilometers of vehicles waited. In addition to Task Force Tarawa's vehicles, there were literally hundreds of vehicles belonging to RCT 1. It was still a particularly lucrative target for the Iraqis. But not unlike his counterpart in RCT 2, the commander of RCT 1, Colonel Dowdy, was reluctant to try to squeeze his regiment through

Ambush Alley. And behind Colonel Dowdy a seemingly endless convoy of supply trucks was waiting to move north.[247]*

This was the kind of evolution that attracted more attention than most subordinate commanders want. On 24 March there were a number of generals on site: Task Force Tarawa's General Natonski; the assistant 1st Marine Division commander, Brigadier General John F. Kelly, whom General Mattis tasked with roaming the battlefield to help him maintain situational awareness; and General Conway, the I MEF commander. Even a retired general was nearby, Major General Ray L. Smith, traveling with the division to gather material for a book, *The March Up*.**

The kind of general who made it a point to see for himself how the fight was going, Conway traveled by helicopter from Jalibah, where I MEF Main was by now well established, to An Nasiriyah. On the way, he flew over what looked to him like a great deal of Marine combat power stretched out on the road, and he remembered thinking, with growing frustration, there should be no holding all of that power back. This made him all the more determined to deliver his message in no uncertain terms. After landing, Conway spent more than an hour at Task Force Tarawa's command post. Then he and General Natonski drove forward in a "soft-skinned" humvee, the vehicle that had replaced the jeep, through fires so intense that three Marines around them were injured. According to one account, the two generals talked matters over with Brigadier General Kelly and Colonel Dowdy while AK-47 rounds snapped overhead. General Natonski remembered later the topics discussed were whether the Marines could hold the bridges and whether 1st Marines could pass through Task Force Tarawa and over the bridges without delay. Throughout, Conway's basic message was simple, find a way to get things moving again.[248]

Word spread wide about the fighting in An Nasiriyah, along with reports and rumors of heavy American casualties. Recordings of a disturbing Iraqi television broadcast showing the killed and captured soldiers made the rounds while corpsmen and doctors waited for the Marine wounded. There was a perception that U.S. forces had suffered a setback and that the war was not going according to plan, especially among the "experts" on television with their nonstop stream of commentary and free advice, usually from thousands of miles away. They were not, almost needless to say, making themselves popular with commanders in Iraq and Kuwait. Reflecting the views of many, the field historian with Task Force Tarawa wrote on 23 March that "[a]ny hopes we may have had for an easy entry into An Nasiriyah, and any larger hopes for a campaign as a series of capitulations, have ended today." The war could be longer and harder than anyone had expected or hoped.[249]

Underlying this new perception was the nature of the fight at An Nasiriyah, which General McKiernan characterized simply as "a damned tough urban fight."[250] It was that, and more. The general expectation had been that the Iraqi soldiers in the regular army divisions stationed in the south of the country would surrender in droves once the Coalition crossed into Iraq, and that the population, at least in the south, where there was a Shia majority hostile to Saddam Hussein's Sunni ruling class, would welcome the Coalition as liberators. But the number of prisoners had been measured in the hundreds, not the expected thousands. This did not mean that the Iraqi Army was fighting hard. On the contrary, it seemed to be simply melting away. What was more surprising was that the irregular forces, especially the loosely organized *Saddam Fedayeen*, literally the "men of sacrifice," soon to be renamed "regime death squads" by Pentagon edict, were willing to stand and fight. Typically in civilian clothes, they were hard to pick out from innocent civilians, whom they were often more than willing to use as human shields or to sacrifice in other ways. There were also numerous reports that they were willing to feign surrender and then open fire on anyone who advanced to take them prisoner.

Speaking about An Nasiriyah, General Conway said that I MEF was facing "hard little knots of Fedayeen."[251] General McKiernan characterized the enemy as "a combination of several different sources, Fedayeen, Special RG, some military, regular army that . . . took off their uniforms . . . but it was a pretty determined enemy."[252] General Mattis spoke for many when he declared that the *Fedayeen* "lack any kind of courage. They literally hide behind women and children, holding them in their houses as they

*The field historian on site noted "the traffic was really snarled around an intersection of 2 major roads" to the south of An Nasiriyah. This was almost certainly a reference to the point where the road from the south branched, with one branch leading to An Nasiriyah and the other to the western bridge, which meant the traffic jam was miles long. (Col Reed R. Bonadonna, "Field History Journal," entry for 23Mar03)

**Gen Ray Smith and his coauthor, Bing West, recorded many interesting vignettes for posterity, including the mood of a young infantryman on the line at An Nasiriyah who seemed to think that events had slowed because of the number of generals there. It is not clear whether the Marine knew that he was talking to a general.

fire . . . They really lack manhood. They are violating every sense of decency. They are as worthless an example of men as we have ever fought."[253] General Amos was equally outraged, commenting later that An Nasiriyah was a turning point for him: "When the Saddam Fedayeen came down and . . . were picking off our Marines, they became, in my mind, cannibals. And my whole perspective on how we were going to fight this war changed."[254]

It is difficult to escape the conclusion that An Nasiriyah was a turning point for many, if not most, Marines. Not only was it the first heavy dose of combat for the Marines, but many things did not go as planned or hoped. The enemy was different than expected, more tenacious and committed, and he was having a certain degree of success. For him An Nasiriyah was a target-rich environment and a large number of Marines needed to pass through a relatively small area. For the Marines it came at a time when many were close to exhaustion, and the battle saw its share of misjudgments, mistakes, and bad luck. All things considered, it is not surprising that there was congestion and confusion, and that the pace slowed.

What is just as obvious is that no one in I MEF gave way under the pressure and that the Marines quickly recovered from the first day of battle in An Nasiriyah. It did not change the force's focus or slow operations for more than a day. On the contrary, there was a hardening of resolve among many Marines. For his part, General Conway did not lose his focus. He consistently pushed division and wing to move north in order to defeat the *Republican Guard's Baghdad* and *Al Nida Divisions*, which lay on I MEF's route to Baghdad, while General Natonski and Tarawa dealt with the *Fedayeen* and other threats in the south.

Even though the firefights, sometimes heavy, continued in An Nasiriyah for a few days, RCT 1 had pushed through the eastern part of the city and started up Route 7 by the afternoon of 25 March. General Mattis' intent was for RCT 1 to move quickly to the north, in the direction of Al Kut, in order to fix the *Baghdad Division* in place in the center of the country long enough for him to get the rest of the 1st Marine Division behind the enemy division, while blocking any Iraqi forces that might attack the Marines from the east. He did not want the Iraqi infantry to be able to fall back into Baghdad.[255]

Although the fighting in An Nasiriyah did not slow the division by much more than 12 hours, the weather did succeed where the enemy had failed. On the night of 24-25 March, the "mother of all sandstorms" moved into the theater, with high winds stirring up massive clouds of sand and slowing operations to a crawl. General Usher, the 1st Force Service Support Group commander, called it the worst sandstorm in 20 years.[256] In the words of the division's command chronology, "Marines choked on the dust and visibility was reduced to almost nothing. Soon, it was blowing so hard that it was difficult to breathe outside."[257] Most air was grounded, leaving fire support to artillery and mortars. But some pilots braved the weather anyway. On 25 March, a field historian watched "a breathtaking performance by two Hueys . . . trying to deliver ammo. They flew straight up the road at about the level of the telephone lines. Between the wind and the prop wash, visibility must have been less than zero. . . . [T]he last I saw of them, they were flying [away] into the dust clouds."[258] Like the Huey pilots, Colonel Steven Hummer of the 7th Marines was not ready to let the weather stop him or his regiment. He himself became a ground guide, personally leading his Marines north in the storm, with "connecting files" behind him, Marines walking between vehicles, literally holding on to the one in front and the one in back, guiding them slowly forward. But finally it was too much even for Colonel Hummer, and he stopped his Marines for the night, putting them in a defensive posture.[259] Hunkering down for the night did not necessarily mean the Marines were safe. In one of the tragic, incomprehensible accidents that occur during wartime, the executive officer of 3d Battalion, 5th Marines, Major Kevin G. Nave, was killed by an earthmover while he was sleeping during the dust storm. A few miles to the southeast at Jalibah, the force was trying to operate out of the "Bug," its expensive custom-made air-conditioned canvas command post. But on the 25th, the lights were flickering on and off and the canvas was flapping vigorously. Marine expeditionary force officers were worried that the Bug would literally blow away, and they took the precaution of passing control back to the rear at Camp Commando for a day.[260]

The sandstorm was followed by thunderstorms, which cleared the air somewhat but created mud, both on the ground and in the air. More than one Marine commented that when the thunderstorm hit, it seemed like it was raining mud; the rain hit the dust suspended in the air and drove it to the ground in wet, heavy drops. Many division Marines went to sleep in sand and woke up in mud on 26 March. For at least a day after the storm, there was a massive cloud of sand over Kuwait, which limited visibility, continued to keep many aircraft on the ground, and made it seem that the sun was shining through a dense, bright yellow filter.[261]

The Washington Post reported that now, in the wake of An Nasiriyah and the sandstorm, "some senior U.S. military officers" were convinced that the war would last for months and would require "considerably more combat power." The United States had kicked in the door, and the house had not collapsed; on the contrary, it seemed to be holding up fairly well in some ways.[262] General McKiernan was concerned not so much about the situation as about what his commanders might be thinking: "The going was a little tough. The Fedayeen . . . and the urban defenses were something we were going to have to deal with. The weather was bad, and we had extended our supply lines." As a result, he felt the need to fly up to see the V Corps commander, Lieutenant General William S. Wallace, to look him in the eye and to "know that we both saw the way ahead," which was Baghdad.[263]

Chapter 6

Toward the Enemy Center of Gravity: Ad Diwaniyah, Al Kut, and the Pause

Even before the storm had completely passed, Regimental Combat Teams 5 and 7 resumed their progress up Route 1, the northwest-southeast axis running to Baghdad, which was by turns a new four-lane highway and a roadbed under construction. On the way, they faced roadside ambushes by a variety of enemy formations that had prepared positions alongside the highway. An incident on 25 March that would ultimately mean a Navy Cross Medal for First Lieutenant Brian R. Chontosh conveys a clear sense of the nature of the fighting. In the early morning hours of the day, Lieutenant Chontosh, an energetic, down-to-earth bodybuilder, who started his career in the Marine Corps on the enlisted side of the house and still shaved his head, was in the lead vehicle of his combined antiarmor platoon, behind four M1A1 Abrams tanks as 3d Battalion, 5th Marines, pushed north toward Ad Diwaniyah, a city some 100 miles south of Baghdad that had been the home of a large Iraqi Army garrison before the war. Suddenly the enemy, described as a mix of irregular and conventional forces, sprang an ambush from the berms on both sides of the highway. The enemy fire struck one of the platoon's vehicles, killing one Marine and wounding another. Lieutenant Chontosh wanted to move the platoon out of the kill zone, which was difficult because there were vehicles both in front of him and behind him. Noticing a break in the berm, he directed his driver to head through it and into a trench filled with enemy soldiers. Once in the trench, Lieutenant Chontosh jumped out, engaging the enemy with an M16 rifle and then with a 9mm pistol until he ran out of ammunition. Then, in the words of the summary of action for his award:

> [H]e . . . grabbed an enemy AK-47 [rifle] and

A convoy of Marines with Regimental Combat Team 5, watched over by scout helicopters, traverses the desert of central Iraq. The combat team had moved hundreds of miles and confronted countless ambushes in the first two weeks of operations in Iraq.

Photo courtesy of Defend America

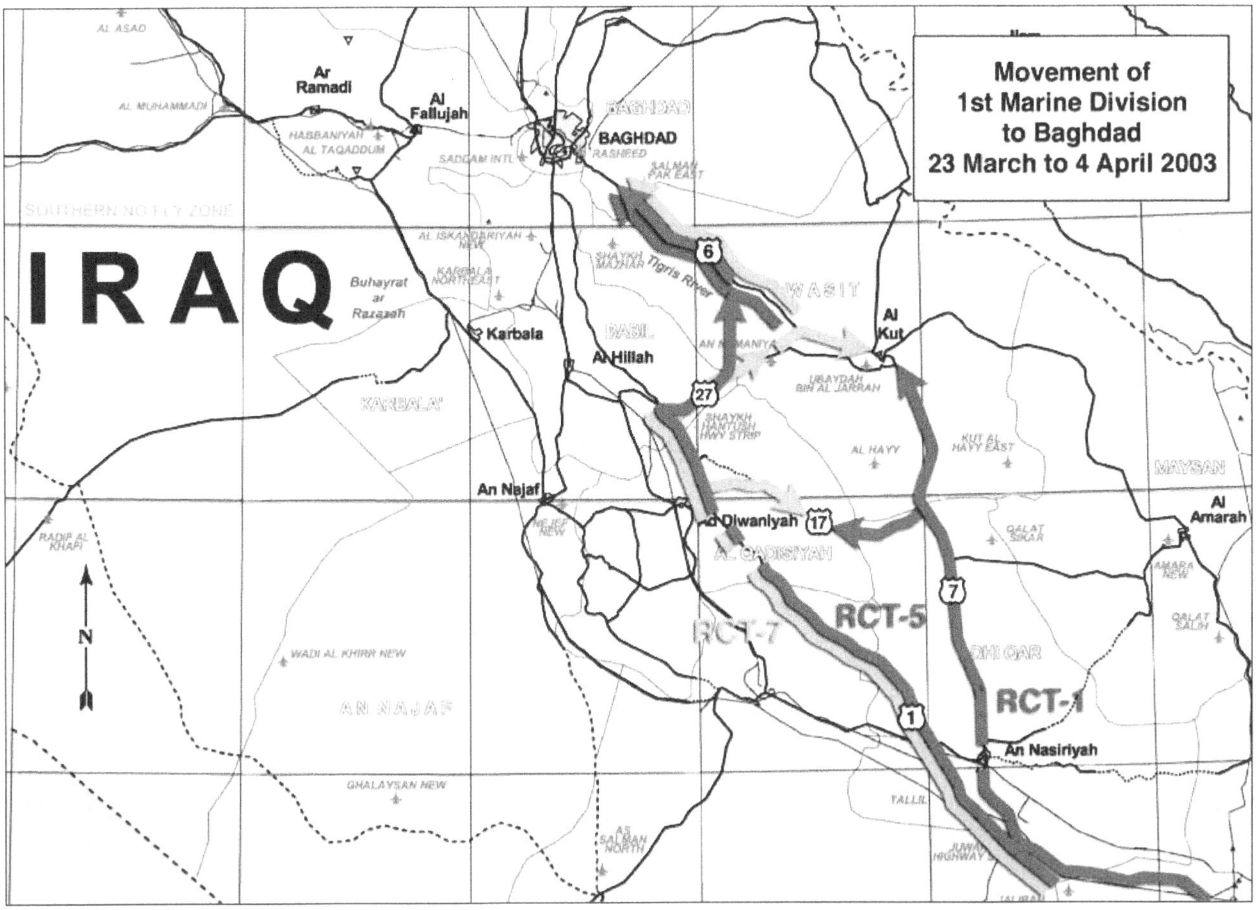

continued to engage enemy soldiers as he continued the attack to clear the trench. . . . When the AK-47 was out of ammo he grabbed another and continued to engage [the] enemy both in and out of the trench under heavy enemy fire. A Marine following him found an enemy RPG [rocket propelled grenade] and gave it to Lieutenant Chontosh who . . . used it to engage a group of enemy soldiers, eliminating the . . . threat . . . His aggressive, violent action . . . undoubtedly saved the lives of many Marines along Highway 1 that day.[264]

Between 26 and 28 March, RCT 5 proceeded to crush *Fedayeen* opposition in and around Ad Diwaniyah. Even before the fighting had ended in Ad Diwaniyah, one of RCT 5's battalions seized the airfield at Hantush after what has been described as "a fierce firefight." Hantush was some 15 miles to the north of Ad Diwaniyah, located on Highway 27, an east-west axis that General Mattis intended to use in the coming days to approach Baghdad from the east. In the meantime, RCT 1 was fighting its way up Route 7 through towns and villages to the junction with Route 17, which was at roughly the same latitude as the city of Al Amarah, not far from the border with Iran, the home of the Iraqi *10th Armored Division*, which was another potential threat, from the southeast, to Marines advancing on Baghdad.

The 1st Marine Division was moving ahead despite the increasing distance from its base in Kuwait. General Mattis had been prepared to rely on organic supplies for a few days; he later said that the 1st Marines, when they crossed the river at An Nasiriyah, fully expected to cut their supply lines "and just break loose and head north," relying on emergency resupply by air when necessary, hence the interest in Hantush airfield, and not having to worry about protecting their supply lines back to the rear.[265] This was a reflection of General Mattis' "logistics lite" philosophy he had been inculcating in all of his troops for months, and of the benefits of General Usher's reorganization, the 1st Force Service Support Group having created a direct support structure for the division, especially the combat service support companies designed to move with the regiments. As stated in the division's command chronology, the reorganization "provided [for a] . . . shared situation awareness . . . [which enabled FSSG to] proactively calculate logistical needs and have them out the door before the customer even registered a request."[266] Although statistics are lacking, this approach most likely helped to reduce the demand for consumables and, perhaps as important, contributed to an expeditionary mentality that made his Marines believe they could go the extra mile without extra supplies.

On a larger scale, the Marine Corps supply system had generally kept up, even though it was stretched. As the deputy Force Service Support Group commander, Colonel John L. Sweeney, commented on 24 March, "the plan is evolving" successfully; there had been "no operational pause due to logistics" because of "what those lance corporals are doing out there."[267] In the daily I MEF situation report for the next day, General Conway reflected satisfaction with the group's efforts to date, especially the hose reel system it had laid from Kuwait to Jalibah, a distance of some 70 miles, in order to deliver fuel. The Force Service Support Group Marines had accomplished this feat in less than half the projected time.[268]

There can be little doubt that it had been a challenge to get supplies from the beach to the forward bases of the Force Service Support Group, which quickly sprang up in Iraq behind the advancing Marines at places like Jalibah and Ad Diwaniyah. It turned out that the Army's 377th Theater Support Command had not been able to meet all the needs of both I MEF and V Corps as originally hoped; the 377th was a relatively late arrival in theater, its headquarters not "closing" until March 2003. Partly because of this and partly because of the challenges the campaign would pose for any command planning to move men and equipment over such a long distance, logistics remained a concern for General McKiernan throughout the campaign, and there were some trade-offs. The main effort, V Corps, was typically supplied first, and the Marines were sometimes left to their own devices. Under Brigadier General Lehnert, the "wholesale" Marine Logistics Command had done what it could to bridge the gap between the 377th and the "retail" service support group and, in general, to meet I MEF's needs. After its arrival in country in December 2002, it had worked wonders in its nearly featureless stretch of Kuwaiti desert now known as Tactical Assembly Area Fox, turning it into a vast logistics base. The general and his Marines often had to improvise, which they learned to do well. For example, the logistics command had to contract for 300-tractor-trailers driven by third-country nationals, many of whom were to drive hundreds of miles into Iraq under dangerous conditions, and then find ways to motivate them to continue working under near-combat conditions. One solution was to award "eagle, globe, and anchor" emblems to the more intrepid drivers. As the Marines moved north, I MEF kept asking the logistics command to keep pace, which it did by, in General Lehnert's words: "using every transportations means available including Marine Corps tactical trucks, Army line haul, contracted third country national . . . vehicles, C-130 . . . air delivery, and rotary-wing aircraft."[269] After the war, General Lehnert concluded that he did not "know how much further we could have gone as the culminating point kept moving north. . . . We had every truck and every driver on the road to the limit of their ability. . . . We were always in a surge mode." There were some notable shortages, especially of spare parts, and there was no reserve.[270]

Transportation Support Group vehicles loaded with ammunition, meals ready to eat, fuel, and water are staged for a convoy north. The 1st Force Service Support Group supported all logistics for the Marine expeditionary force in Iraq.

JCCC 030401-M-6910K-034

What the Marine Logistics Command delivered to the 1st Force Service Support Group, the group pushed forward to its frontline customers. General Usher said it was sometimes a matter of "brute force logistics." Despite all the group's careful preparations, sometimes it came down to Marines muscling their way through a problem. When the system was stressed, the general added, "[I]t was not pretty. It was not elegant. It was just sheer adrenaline."[271] This was especially true as the Marine division sprinted toward Baghdad:

> [Resupply] was accomplished by the integrated, rapid distribution of fuel, water, rations, and ammunition to the nearest SA [support area] or RRP [repair replenishment point] to the fight, moved by [FSSG] assets . . . and in some cases [by] the MLC, at distances farther than anyone had imagined prior to the beginning of the war. At the height of the action, more than 250,000 gallons of fuel were moved on a daily basis from as far south as SA Coyote [in Kuwait] to as far north as SA Chesty at the An Numaniyah Airfield, stretching more than 300 miles over improved and unimproved highways [in the

The Missile Attack on CFLCC

Marines and soldiers were increasingly becoming aware of the fact that this was a war without defined front lines and rear areas. Threats could come from any quarter, when least expected. But one week into the war, the rear areas, especially in Kuwait, seemed fairly safe, and the mood at Camp Doha was, if not exactly relaxed, at least moving into a smooth wartime routine. During the first few days of the war, everyone at Doha had worn their chemical suits and carried their gas masks; now the soldiers and Marines had the suits and the gas masks with them but were not wearing them. CFLCC's battle update assessment, which was "the" daily brief in the ultramodern command center, on 27 March, was no exception. It was chaired by General McKiernan, dressed as usual in his freshly starched desert battle dress uniform. The staff was moving crisply through its agenda when the alarm came from one of the air defense liaison officers: "Lightning, lightning, lightning!" which meant that an Iraqi missile launch had been detected and that the target was Kuwait. Everyone in the large amphitheater paused to put on their gas masks, which would have offered some protection in case of a chemical or biological strike, but none against the effects of high explosives. With his gas mask in place, General McKiernan went back to work, speaking through its mouthpiece, his voice calm and only slightly garbled. A few seconds later there were two deep detonations nearby, and then a third and maybe a fourth detonation, the sounds of two outgoing Patriot missiles and of at least one Patriot striking the incoming missile. Bits of debris rained down on Doha as the Giant Voice, for once, late, sounded the alarm. In the battle update assessment the gas masks stayed on for a few more minutes while experts tested for the effects of special weapons. Then it was back to business as usual, as if nothing had happened. But something *had* happened. The incident was captured on Cable News Network (to be shown after the war in a documentary on CFLCC's "War Room") and was analyzed and chewed over among the men and women at Doha for a few days. One German artillery officer, attached to C/JTF-CM, claimed to have learned from a good source that the Iraqi missiles fired at Kuwait had not been fired by a fire direction center; instead, the Iraqis were using a form of dead reckoning, calculating the distance and direction to a well-known target like Doha, and simply cranking in the right numbers. If so, they had done an excellent job. The computers at the Patriot missile battery showed that the trajectory of the missile, probably an "Ababil" missile, was such that it would have struck the command center itself or the building next door, which could have wiped out the CFLCC command group. That was not quite the end of it. In this war, neither CFLCC nor any other headquarters, especially in the rear, was a place where anyone showed much emotion of any kind. Maybe it was all the technology that made emotion seem out of place. Most soldiers and Marines were bone-tired, stretched close to their limits by impossibly long and stressful days. On the staffs, some never forgot that what they did, or did not do, could mean life or death for someone a few hundred miles away and pushed themselves even harder. But few ever seemed particularly happy or sad, except for the time when a few days later, during another battle update assessment, the officers in the command center offered the Patriot "missileers" a spontaneous round of applause.*

*MajGen Robert R. Blackman intvw, 31May03 (MCHC, Quantico, VA); Reynolds, Journal, entries for 27Mar-8Apr03; CNN Presents, "Inside the War Room" (Atlanta, GA: CNN DVD, 2003); Maj Robert K. Casey intvw, 27Apr03 (MCHC, Quantico, VA); Fontenot, et al., *On Point*, p. 98; Franks, *American Soldier*, pp. 506-507.

DVIC DM-SD-04-06145

Cpl Alvin Hicks of Marine Wing Support Squadron 373's bulk fuel section refuels an AH-1W Cobra helicopter gunship from the 3d Marine Aircraft Wing's forward refueling point on Jalibah Air Base, Iraq.

face of] . . . the unconventional threat along the . . . MSRs [main supply routes].[272]

The bottom line is that despite some serious challenges, the Marines of the group and the logistics command consistently managed to keep supply ahead of demand on the battlefield and thereby enabled operational success.

That helps to explain why, for many Marines, what happened next was as unexpected as many of the other challenges in this campaign, CFLCC ordered a halt in the march toward Baghdad. According to the I MEF situation report for 27 March, it received a CFLCC order to halt the attack north and focus on securing its lines of communication. The force in turn passed the order to division. General Mattis "did not want the pause. Nothing was holding us up." Moreover, he believed that his troops, especially 5th Marines in Hantush, were in an exposed position, and he did not want to leave them in what amounted to a holding pattern. As a result, he felt constrained to give "one of the toughest orders [he] . . . ever had to give to an assault battalion that had taken ground [and] lost men doing it." He told them to withdraw to a more defensible position, which they did despite their infantryman's "So, now what do they want us to do?" reaction to an order that didn't make sense to them.[273*]

Underlying the order were CFLCC concerns about resupply, which had apparently percolated up from V Corps; during the sandstorm, some Army units had nearly exhausted their supplies. The obvious corollary was now, with the *Fedayeen* threat, who would protect the force's ever-longer supply lines—the long, dusty, slow-moving convoys from Kuwait to the front lines? There may be some validity to the argument that this thinking had its roots in the various prewar discussions about a lengthy "operational pause" between the opening phase of the war and the fight for Baghdad in order to build up supplies and reinforcements. After the war, some Marine operators talked about a link between the prewar discussions and the wartime pause.[274] While I MEF consistently opposed a lengthy operational pause and division opposed any kind of a pause, General Conway opted for the middle ground: a brief pause for I MEF to catch its breath, not to mention staying in synch with the Army corps on his flank, which

*A related issue was that Mattis wanted the Iraqis to continue to think he intended to come at Baghdad from Route 1. If they focused on Hantush, they might realize he was planning a lateral move before threatening the capital from the east.

DVIC DM-SD-05-04645

An aerial view of the base camp Task Force Tarawa established near An Nasiriyah in preparation for the attack toward the Iraqi capital following the pause.

was, after all, still the main effort. On 25 March, for example, he had discussed the situation with his staff, saying that the enemy attacks "will effect . . . our CSS convoys. These huge long supply lines are a problem. . . . Rear area security continues to increase in importance. . . . You might need to look at pulling combat power in order to secure the key areas."[275*]

One of the threshold issues was just how long the pause would last, and what effect it would have on the plan. On 28 March, General McKiernan flew from his headquarters in Kuwait to Jalibah, because he wanted to meet face-to-face with his two corps commanders, Generals Conway and Wallace, to discuss the situation and its implications for the upcoming attack on Baghdad, the next phase of the war. For General McKiernan, two enemy centers of gravity were now Baghdad and the Iraqi paramilitary forces, which could impede further progress to the north.

Accounts of the conference vary somewhat, but there is general agreement that both commanders talked about the threats they wanted to address before moving closer to Baghdad.[276] Wallace said he needed time to position his corps. Conway mentioned what sounded like tasks he wanted to accomplish before the attack on Baghdad: I MEF was committed to "a systematic reduction of the bad guys in An Nasiriyah," a reference to the ongoing fight for that city to secure I MEF's rear; the British division needed to execute some "pinpoint armor strikes," that is, raids into Basrah. Referring to RCT 1, he commented that "Joe Dowdy was in a 270-degree fight" on Highway 7 as his command made its way north through the heart of the country, occasionally encountering stiff resistance. General Conway later remembered making the case for a pause of approximately three days.

The upshot of the conference was a relatively open-ended decision by McKiernan for both I MEF and V Corps to "take time to clean up . . . before we commit . . . to the Baghdad fight, because once we commit to the Baghdad fight, we cannot stop."[277] There would be a pause, a chance for securing the rear areas and for supplies to catch up, which, if all went well, would be relatively short, no more than "several days." It would not be a lengthy operational pause to wait for heavy reinforcements, and it would not keep CFLCC from the main event. When he spoke to his staff later on the same day, General Conway downplayed the pause and stressed that the focus remained on Baghdad.[278]

As the operational pause began, I MEF shifted its

*The Marine general speaking out against the pause was Gen Mattis. Gen Conway also discussed the issue during an interview for the History Channel series on the Iraq War (released in 2003), saying he had welcomed the opportunity to do some consolidating and resupplying. *The Washington Post* reported a video teleconference on 25 March among Gens McKiernan, Wallace, and Conway, with McKiernan soliciting recommendations, Wallace expressing concerns, and Conway wanting to continue on to Baghdad. There is a lengthy discussion of the "pause" in West and Smith, *March Up,* which conveys a picture of hard-charging Marines being held back because the Army was worried about its supply lines. (Rick Atkinson, Peter Baker, and Thomas E. Ricks, "Confused Start, Decisive End," *The Washington Post*, 13Apr03, p. A-1; Bing West and MajGen Ray L. Smith, *The March Up* [New York, NY: Putnam, 2003], pp. 73, 84)

Photo courtesy of CFLCC

LtGen James T. Conway, left, and MajGen James N. Mattis discuss preparations for the final phase of the war. The capture of Baghdad would not mean an end to hostilities but a shift to a post-war phase.

focus somewhat, attacking pockets of resistance in its area of operations while keeping its eye on the ultimate goal. Fragmentary Order 040-03 outlined the missions of defeating "paramilitary forces in zone [in order to] protect MEF lines of communication and set conditions for continued attacks north to defeat the Baghdad and Al Nida [*Republican Guard*] divisions."[279] General Conway and his staff seem to have made it a point even to avoid the word "pause"; they preferred to talk about "throwing elbows" in a different direction. When on Saturday, 29 March, the general quipped, "Enjoy your Saturday night, kick back and relax, and we will tell you when the war starts back up again," he did not intend for anyone to take him seriously.[280] Nor did he want anyone to think I MEF had shifted to conducting counter guerrilla operations. In his view, I MEF was simply "re-cocking" for the next phase: attacking pockets of enemy resistance, shoring up logistics bases, building air bases, giving the 3d Marine Aircraft Wing time to conduct the "shaping" operations that had not been possible earlier. As General Conway remarked on 30

The Battle of the Icons

The 3d Marine Aircraft Wing's "kinetic" and "non-kinetic" effects, shorthand for bombs and leaflets, on the Iraqi *10th Armored Division* were so powerful that it was soon little more than an icon on U.S. computer screens. I Marine Expeditionary Force had monitored the situation and was satisfied that the *10th* was no longer a threat, but the icons worried higher headquarters, perhaps even someone at the Pentagon, and the order eventually came down from CFLCC for I MEF to "neutralize" that division. And so, in early April, General Conway directed General Natonski to either capture or destroy its remnants. Task Force Tarawa put together a smaller task force from 1st Battalion, 2d Marines, and the 24th MEU (SOC), which advanced some 300 kilometers across Iraq in short order, no small feat in itself, and, though prepared for heavy fighting, found nothing but cheering crowds when they drove into the flat, dusty town of Al Amarah on the Tigris. The attack spanned 8-10 April.*

*Col Jeffrey Acosta, "OIF Field History Journal," 2003, entry for 20May03 (MCHC, Quantico, VA); 2d MEB ComdC, Jan-Jun03 (GRC, Quantico, VA); Michael Wilson, "Two Marine Battalions Turn to Confront the Remnants of a Lurking Iraqi Division," *The New York Times*, 8Apr03, p. B-8.

May, "while we were stationary [on the ground], we were in fact attacking with our air," putting out 300 to 320 sorties per day on the enemy.[281] The *Baghdad Division*, arrayed around Al Kut, received constant pressure from the wing, as did the *10th Armored Division* to the south in the vicinity of Al Amarah. Both were ultimately degraded virtually to the point of ineffectiveness by the air attacks.

The 1st Marine Division did not hide its impatience to move on toward Baghdad. Again and again, in its formal and informal communications, it spread the message that it was "anxious to resume the attack . . . [because] the best way to secure our locs [lines of communication] is to rapidly move n[orth] to collapse the regime."[282] General Mattis himself penned the comment on 29 March that he was "convinced that the enemy situation is such that we could cross the Tigris and destroy the *Baghdad Division* without interference from the *10th Armored Division* within the next 72 hours."[283] General Conway was only slightly less forward-leaning, reporting on 30 March that "the conditions for attacking north are rapidly being set and should be in place within 2-3 days." It was his view that I MEF had nearly all the supplies it needed to move forward, and he was hearing that after the continuous air attacks, the enemy simply was not there in large numbers.[284]

On 30 March, General Mattis flew to Jalibah to meet with General Conway and outline his plan for getting things moving again. General Mattis found that he was preaching to the choir, I MEF planners being just as eager as he was. General Conway agreed that the security situation would allow the division to proceed with its plan for a limited objective attack across the Tigris to isolate Al Kut, which in turn would open up a number of further options for both the 1st Marine Division and I MEF. This was true even though Conway believed that Saddam would order an attack by weapons of mass destruction once the Marines reached Al Kut. Chemical warfare "attack likely when attacks on Baghdad resume" were the words of I MEF's situation report for 28 March. "Trigger depends on U.S. success against . . . forces [in the vicinity of] Al Kut."[285]

In the forward command post a few miles south of Ad Diwaniyah, MajGen James N. Mattis, left, meets with the commanding general of the 3d Marine Aircraft Wing, MajGen James F. Amos, to discuss air support for the upcoming offensive.

Photo courtesy of Col Charles J. Quilter II

The I MEF plans were not just about the 1st Marine Division. Task Force Tarawa's status was especially important, since once An Nasiriyah was secure the task force was to "expand its battle space to the north" along Routes 1 and 7 to guard the division's rear as it moved farther north. Between 1 and 6 April, Tarawa focused RCT 2 on this task while continuing its increasingly successful "three block war" in An Nasiriyah—that is, the mix of combat patrols and civil affairs work that was required after the major battles had been fought. Task Force Tarawa had been augmented by 15th MEU (SOC) and 2d Battalion, 25th Marines, which assumed joint responsibility for An Nasiriyah, freeing RCT 2 to operate to the north.[286]

During the preparations for the final phase of the war, the 1st Marine Division and the 3d Marine Aircraft Wing continued to place a premium on face-to-face meetings. The meeting between Generals Amos and Mattis on 31 March is representative. On that morning, General Amos, the former fighter pilot, flew himself and a few members of his staff in a CH-46E Sea Knight helicopter from Kuwait to the division's forward command post alongside Highway 1, a few miles south of Ad Diwaniyah in the center of southern Iraq. Even though the visibility was still poor, and the desert below largely featureless for long stretches, the general flew at about 100 feet on account of the surface-to-air missile threat. It was an exhausting six-hour round trip.

General Mattis met the helicopter when it landed and walked his guests to the command post, which was little more than a few tents and camouflage netting. He began by praising the wing for its support, especially for flying unmanned aerial reconnaissance vehicles for the division. This was like having a small television camera in the sky to scout the terrain just ahead, which, together with the high-flying Navy P-3 Orions that General Mattis had arranged for, delivered an excellent picture of the battlefield. The generals then discussed the upcoming offensive, with General Mattis describing his plan and asking for

Photo courtesy of CFLCC

A burning Iraqi T55 tank, destroyed by the 3d Battalion, 4th Marines, sits along Highway 27 north of An Numaniyah, the site of an enemy military compound.

more support from the wing. Two issues stood out: one was bridging the canals and rivers that stood between the division and Baghdad; the other was resupply. Since the 1st Service Support Group's convoys had to drive literally hundreds of miles to reach the front, the obvious alternative was for the wing to airlift supplies when shortages loomed.

After two hours of talks, General Amos and his staff flew up to Hantush to personally evaluate the suitability of a field-expedient "highway airstrip" for Marine cargo planes. General Mattis, who was immensely grateful for the wing's support and his good relationship with General Amos, quipped later that General Amos showed "more bravery than good judgment" when he flew himself on to Hantush. They "approached by circling over a large date palm grove [which later turned out to have contained enemy fighters] . . . and landed on the superhighway which was lined with [the] vehicles of the 5th Marines," who had moved back into the area a few hours earlier. Amos found the 8,000-foot runway suitable for his pilots even though it was little more than a four-lane highway from which the Marines had recently removed the centerline traffic dividers.[287] That night, flights of heavily loaded KC-130s Hercules aircraft landed at Hantush using night-vision goggles. The division later described Hantush as a "critical logistics hub" for the final push north. Finally, before returning to his headquarters in Kuwait, General Amos stopped briefly at I MEF's forward command post in Jalibah to touch base with General Conway.[288]

Writing about the meeting, one of the officers in General Amos's party, field historian Colonel Charles J. Quilter II, observed how much things had changed since Desert Storm, especially in terms of communications between the actuals: "There was the twice daily . . . video teleconference. They also talked on the phone a lot. That really struck me. Compared to Desert Storm . . . they talked far more than their predecessors [in that conflict]. . . . [They] often talked late . . . [into the night] about . . . the latest developments and what kind of air support the division would need the next day. . . . They were called the 'Talking Jims' by their staffs."[289]

The expeditionary force resumed the offensive to the north on 1 April. With RCT 7 following in trace, RCT 5 advanced northeast along Route 27, seizing a bridge over the Saddam Canal, a continuation of the north-south waterway the Marines had encountered near An Nasiriyah. The next day, RCT 5 seized two crossings over the Tigris, putting it astride a major route that ran between Baghdad and the city of Al Kut, where the British had suffered a disastrous defeat against the Turks in 1916 during World War I. This enabled the division to complete the destruction, in the vicinity of Al Kut, of the *Baghdad Division*, which RCT 1 had fixed in place by advancing from the south, thereby putting it at the mercy of RCT 7 advancing from the northwest. The division history records 3 April as the day RCT 7 destroyed the enemy division's two western brigades.[290]

General Mattis made a point of keeping his troops

from actually going into the city of Al Kut; he did not want to become bogged down in urban combat, and he did not want to assume the responsibility for governing the city, which he could incur under the "Law of War" if he entered the city limits. When he was content that the enemy was surrounded, largely combat ineffective and unlikely to attack, he left Al Kut to its own devices, posting a recon battalion outside the town to make sure the Iraqi commander did not "get brave."[291] The division was now free to turn its attention to Baghdad.

It was in the wake of this attack that General Mattis, in a move that attracted considerable attention and stirred some controversy, relieved Colonel Dowdy as the commanding officer of 1st Marines and replaced him with his operations officer, Colonel John A. Toolan. (Generals Conway and Mattis discussed the relief before it occurred.) Colonel Dowdy, a well-respected Marine who placed a premium on the well-being of his Marines, apparently fell victim to the general's overriding quest for speed. The division spokesman said, simply, "It was a decision based on operating tempo." On 4 April, in a tent at the division's command post, there was a difficult meeting between General Mattis and his assistant division commander, Brigadier General Kelly, on the one hand, and Colonel Dowdy on the other. Dowdy stood on the record of his Marines, who had been fighting their way up-country and getting the job done, perhaps not as quickly as Mattis and Kelly wanted. According to Dowdy's account of the meeting, Mattis told the colonel he was being relieved and asked him to empty his sidearm and turn over his ammunition, which Dowdy said would not be necessary. Before long, Dowdy was on a helicopter to Kuwait. A considerate man not given to undermining his brother officers, Colonel Toolan had not known what was coming but obeyed the order to replace Dowdy. When he arrived at the regiment, he let it be known that he would carry on where his predecessor had left off, that the regiment was bigger than any one man. Colonel Dowdy could not have agreed more, posting a message on an internet website saying he remained loyal to the division and its leaders. He spent the rest of the war serving as an aerial observer in one of the P-3s flying over the battlefield.[292]

Chapter 7

Baghdad: Going Down Fast, Going Down Final

The mantra had always been, Baghdad is the enemy's center of gravity; the purpose of the campaign is to remove the regime, the means to that end is capturing Baghdad. It was where Saddam's power resided, both symbolically and otherwise. CentCom and its subordinates were consistent in their assumption that Saddam had to defend the capital. While he had stationed relatively weak regular army divisions in the south, Saddam had kept his best forces around Baghdad, the *Republican Guard* and the *Special Republican Guard*. The thinking was still that the Iraqis would set up concentric rings of defense around Baghdad. Coalition Forces Land Component Command (CFLCC) maps showed a ring around Baghdad and an area labeled the "Red Zone." The deeper the U.S. forces penetrated into the Red Zone, which took in Al Kut in the east and Karbala in the west, the tougher the fight would become. Supported by fanatic militiamen, retreating Iraqi forces would make their last stand in and around the capital.* The Red Zone also was where Saddam Hussein was most likely to order the use of weapons of mass destruction, especially if the fight went badly for him and he had little or nothing left to lose by way of international support. He would, presumably, prefer to stop the Coalition with the weapons outside his capital. But with Saddam Hussein, who could be sure what the plan was?

Since Baghdad was the one city that CFLCC could not bypass, the question, as early as the summer of 2002, was how best to attack it and turn the nightmares, if not into pleasant dreams, at least into tolerable slices of reality. The V Corps and CFLCC planners, including Colonel Kevin Benson, the plans officer who had solid relations with I MEF planners, took the approach described as "systems-based planning," which had grown out of the work of some very good military theorists at the Army's School of Advanced Military Studies, the Marine Corps University, and the Warfighting Laboratory at Marine Corps Combat Development Command in Quantico. Soon these theorists engaged I MEF planners, who made a contribution of their own, and learned an approach that the Marines could adapt to fit their plans. The idea was to think of a city as a system or, better yet, as a system of systems. On the one hand were the systems used by the regime to control the city, the other kind of "power points" such as the police and the military or the government-run media, not to mention symbols of power such as palaces. On the other hand were the systems that made the city run—water, electricity, and transportation. There was of course some overlap between the two categories. An airport, for example, could fit into two or three categories. By analyzing a city, planners could map the relationships between systems and identify the "key nodes." These could be attacked with precision-guided weapons from the air or with raids on the ground. Then there would be no need for costly house-to-house, block-to-block fighting, let alone the wholesale destruction of the infrastructure, which the Coalition wanted to preserve for the postwar phase.[293]

To turn theory into practice, conferences and seminars on urban warfare were held in late 2002, like the ones at CFLCC in Kuwait in December.[294] The Army, and CFLCC, came to favor a concept of operations with two basic steps. The first was encircling and isolating Baghdad. This would prevent reinforcements from entering the city and keep prominent members of the regime, especially Saddam Hussein and his sons, from escaping. The second step was to establish bases outside the city limits and then conduct "in and out" armored raids to attrite the enemy. Supported by attack helicopters, with fixed-wing support on station nearby, the raiders would identify points of resistance, hit them "hard and quick," then get out or simply advance along a particular axis and destroy whatever opposition presented itself. In December 2002 General David McKiernan summed up the plan as one to "isolate

*There might even be scenes like that portrayed in the movie *Blackhawk Down*, about U.S. soldiers mired in a fight in 1993 in downtown Mogadishu, where U.S. technology was hard put to overcome paramilitary fighters in a warren of alleyways and ruined buildings. It was the kind of place where helicopters became vulnerable to rocket-propelled grenades fired by young men and little boys, and elite troops lost their way and their lives. The vivid images from that movie, based on the gripping and carefully researched book by journalist Mark Bowden, were familiar to virtually every American soldier and Marine in the Iraq War. Those images were the stuff of nightmares for planners and commanders.

Photo courtesy of Defend America

LtGen James T. Conway, left, Col Steven A. Hummer, center, the 7th Marines' commanding officer, and Col Larry K. Brown, operations officer for I Marine Expeditionary Force, meet to discuss the Baghdad offensive.

Baghdad, establish an outer cordon which controls movement in and out of the city, and then a series of forward operating bases . . . to . . . attack . . . specific targets in the city [and return to base,] . . . or [to] seize and secure specific targets." The process, which could be lengthy, would continue until the enemy was too weak to oppose an occupation.[295*]

The CFLCC plan made sense for a force made up of mechanized infantry; the 3d Infantry Division, for example, had many tactical vehicles at its disposal, but surprisingly few "dismounts"—1,200 to 1,600—infantrymen who were trained to fight on foot. Baghdad was an open city in the sense of having broad boulevards leading in and out of town that were not too bad for armor. This made for an exception to the Army's doctrinal reluctance to use heavy armor in urban areas. The Army had tried the concept on a limited scale in the city of Najaf on the way to Baghdad, and the results were not inconsistent with the British example in Basrah. From their base at the airport outside the city, the UK division was staging carefully planned raids against specific targets. Intelligence collection on the streets of Basrah drove some of the targeting, a modus operandi that the British had learned the hard way in Northern Ireland over the preceding thirty years. Compared to the Americans, the British tended to use less armor and more infantry, and to drop off snipers and spotters when the main body withdrew. By early April, they appeared to be enjoying a modicum of success.[296] Their success was contrasted with perceptions of Marine problems in An Nasiriyah. The comparison was not apt, as the goals of the two operations were different. The Marines needed to go into An Nasiriyah to get to the other side of the river in a hurry, while the British goal was to secure Basrah when the time was right. But at least for some, the perception was there. Planners at various Army, Marine, and joint headquarters were aware of what was happening in Basrah, and while it is impossible to pinpoint an instance when anyone copied a particular British tactic, the British approach seemed to confirm some of what the planners had been saying about urban tactics.[297]

The 1st Marine Division concept was different both from the CFLCC/Army concept and from the British concept, not surprisingly, since it was a more traditional infantry division, with some 6,000 rifle-

*Rick Atkinson in his book, *In the Company of Soldiers*, notes that this was a departure from the Army's reluctance to commit armor to an urban area and he also notes Gen Wallace's preference for staying out of the cities and defeating the Iraqi Army and the regime on other ground. This was obviously a minority opinion. Some Army planners believed the in-and-out raids could take weeks. (Atkinson, *In the Company of Soldiers*, pp. 185-186, 218, 287-288)

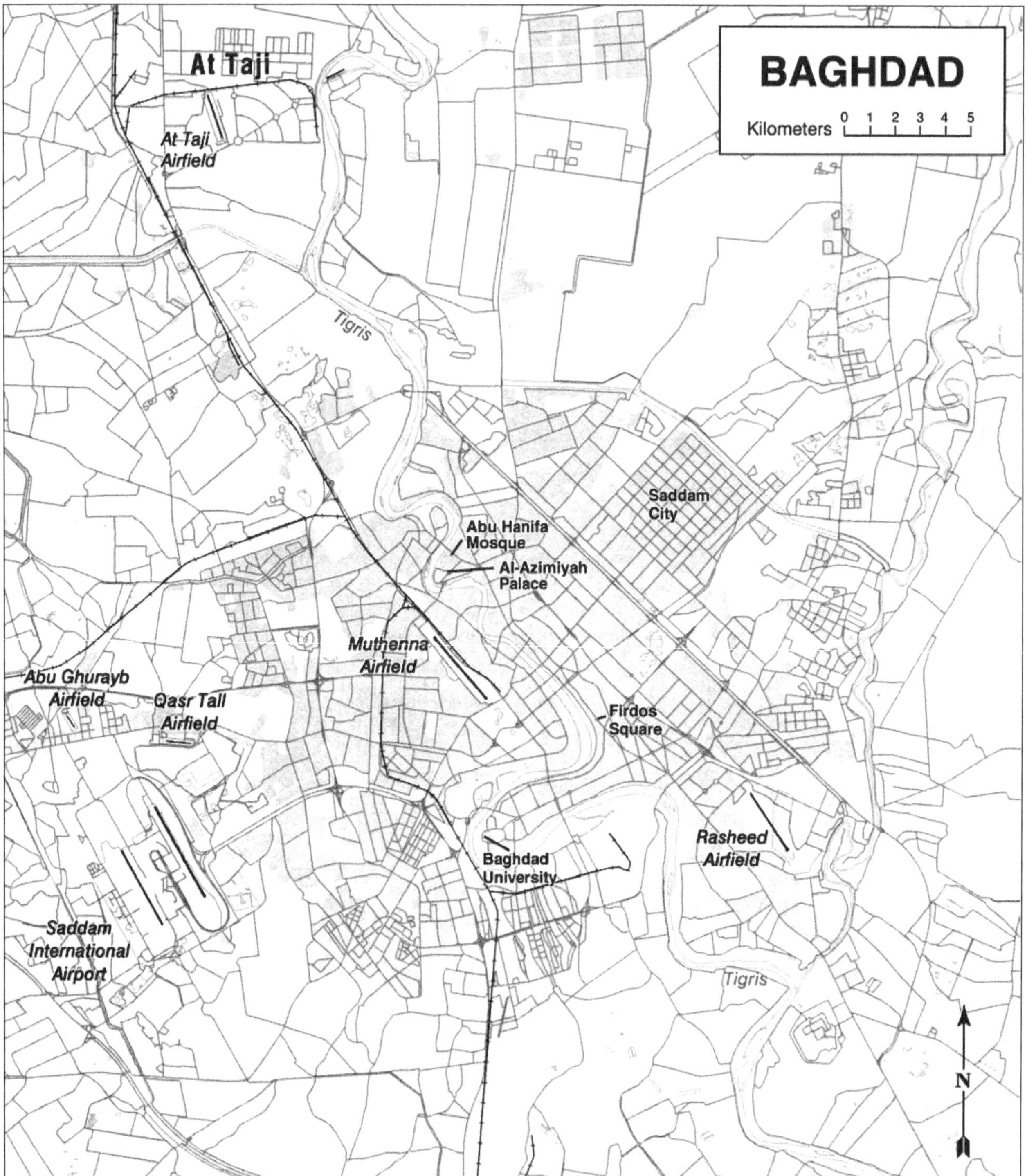

men. As early as 1999, the Corps' Warfighting Laboratory had used division forces to run the "Urban Warrior Advanced Warfighting Experiment," exploring some of the problems Marines would face in cities in the 21st century. In California, in December 2002, division had conducted its own seminars and training on urban warfare, spending three days talking about how it would fight in Baghdad. Among the attendees were representatives of the Warfighting Laboratory, Marine Aviation Weapons and Tactics Squadron (the Marine equivalent of the legendary Navy's "Top Gun" School), I MEF, and the 3d Marine Aircraft Wing. The seminar was followed by an "Urban Combined Arms Exercise" at the abandoned airbase at Victorville. The exercise ran two battalions through the kinds of challenges they might face in

Photo courtesy of CFLCC

The air traffic tower of Saddam International Airport, soon to be renamed Baghdad International, looms over the terminal. The Coalition placed considerable emphasis on seizing the airport, both as a symbol and a useful piece of real estate.

Baghdad, including "militias competing for power on the streets, the breakdown of civilian authority, unruly crowds at food distribution centers, car bombs, and snipers hiding in crowds."[298] The upshot was that division did not like the idea of seizing objectives, giving them up, and then having to seize them again later. "Withdrawals from portions of the city after seizing raid objectives would embolden the enemy and lessen the 'dominating effect' the division wanted to portray to the enemy and to the international media." Moreover, "identifying important targets by raiding and then abandoning them would give the Iraqi fighters the opportunity to reoccupy, mine, booby trap, or preplan fires."[299] In a postwar interview, General James Mattis said: "We were not eager to set up . . . bases around the . . . city and raid into it and back out at any point."[300]*

The division staff followed its commander's thinking about raids. First Mattis directed his intelligence officers to prepare a list of target packages, worthy objectives in the eastern half of the city, which ranged from military installations to media centers to government offices that could be raided or seized and held. The list kept on growing. As the list grew, it seemed to make less and less sense to think of the operation in terms of "in-and-out" raids. Why seize one site and then give it up only to return the next day to seize a nearby site? Ultimately, the division began to consider breaking the city into zones and assigning them to the maneuver regiments, and even to its artillery regiment, 11th Marines.[301]

What the division or I MEF thought about how to run the urban fight did not matter as much as what two other commands thought about the city. After all, CFLCC had long since "assigned" Baghdad to the Army's V Corps to avoid the problems that came with having two corps-level commands responsible for the same objective.[302] In late 2002 it seemed that V Corps would be able to split the city between the 3d and 4th Mechanized Infantry Divisions while the Marines helped to maintain the cordon around it. But then after the Turks refused to allow the 4th to pass through their country into Iraq, CFLCC had considered using Marine units in place of that division. One proposal was to put one or more Marine regimental combat teams under V Corps' tactical control. A version of this proposal surfaced in a CFLCC draft of the order for Baghdad as late as 25 March. This did not sit well with Marine planners, who argued that a Marine regiment did not have the kind of robust communications suite needed to communicate with a corps-level headquarters, and that the best way to employ Marines was as a Marine air-ground task force. Even without a boundary shift, that is, even if Baghdad proper were the province of the Army, and the Marines stayed outside the city limits, an intact air-ground task force could support V Corps more effectively, especially if given some latitude to control its own operations. "How to fight the MAGTF" was one lesson that seemed to be on the curriculum every semester.[303]

General McKiernan gave fair consideration to General James Conway's arguments but did not come to a final decision about Baghdad before G-Day.

*This was not only division policy, but it was also the conviction of at least one regimental commander, Col Steven Hummer of 7th Marines, who did not want to withdraw from hard-won gains. (Col Steven A. Hummer intvw, 13Feb04 [MCHC, Quantico, VA])

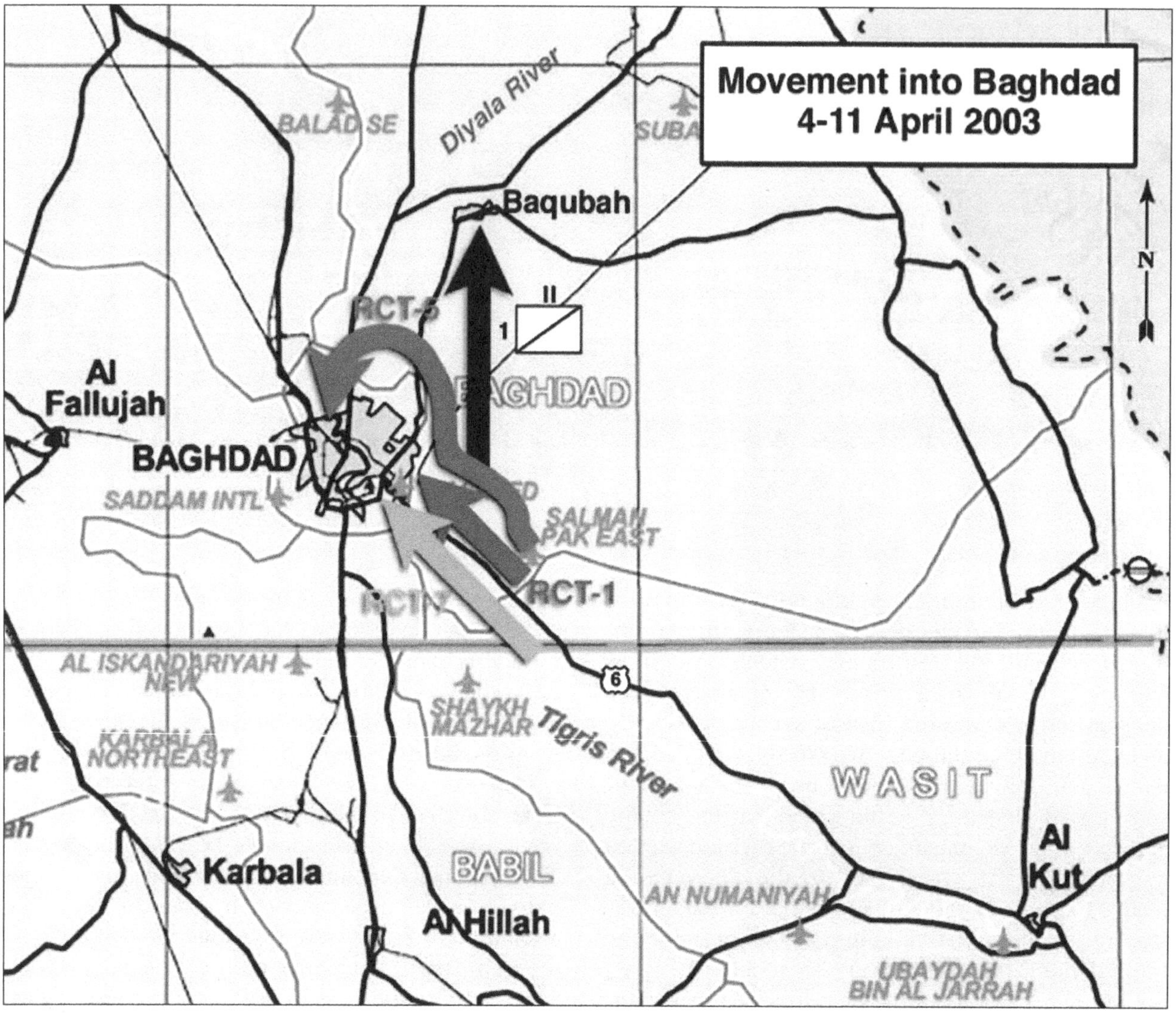

McKiernan had approved a branch plan to bring forces into Baghdad from both I MEF and V Corps if it made sense to do so.[304] Earlier, he had said, "[I]f there is a decision . . . about introducing I MEF forces into Baghdad . . . I will establish the boundary . . . [which] would logically be . . . the Tigris." The assumption was clearly that I MEF would come from the east and V Corps from the west.[305] In the words of Marine planners, even after "the battle [for Iraq] began, the issue continued to evolve, and, as the Coalition neared Baghdad, the decision was made to [ap] portion the city along the Tigris River," assigning the eastern half of the city to the Marines.[306]

It appears that General McKiernan made that decision in early April when the "Baghdad fight" was finally taking shape, and he was ready to predict that the regime was "going doing fast, going down final." By 3 April, his staff had issued CFLCC Fragmentary Order 124, which delineated the boundary between I MEF and V Corps. This would, McKiernan emphasized, be a "coordinated two-direction attack with I MEF attacking to seize [an intermediate objective] . . . and then beginning to work into Baghdad from the southeast."[307]

Was it a last-minute decision, one that had to be made earlier than expected? The watchword for the campaign had always been speed. Few had thought that CFLCC troops would be on the outskirts of Baghdad in early April. Even a forward-leaning officer like General James Amos thought it would take some 55 days just to get to Baghdad.[308] Did the quick tempo outstrip prewar plans or, more precisely, the planning process? This was an issue that General McKiernan had been thinking about for months. He had a base plan, one with branches, designed to allow him the flexibility to change circumstances that no one could predict in advance. In that sense, everything was going according to plan.

While 1st Marine Division had been reducing the *Baghdad Division* near Al Kut, the Army's 3d Infantry Division, with the 101st Airborne Division in trace, had continued its march up through the western desert, reducing resistance in and around the cities of Najaf, Karbala, and As Samawah. The relationship between these two divisions was not unlike the relationship between 1st Marine Division and Task Force Tarawa, with the heavier force in the lead and the lighter force securing the lines of communication. By 3 April, the 3d was on the outskirts of Baghdad, ready to move on Saddam International Airport, soon to be renamed Baghdad International Airport, or "BIAP," as it came to be known to most American military personnel. (General McKiernan viewed the airport as one of the enemy system's key nodes, both as a symbol and as a useful piece of real estate; he placed considerable emphasis on seizing it and planting his own flag on it.) On the same day, the Marine division was once again on the march toward Baghdad, from the southeast. Along the way, the 5th Marines fought one of the fiercest engagements of the war on Route 6 near the town of Al Aziziyah.

The enemy was a mix of *Republican Guard* troops, *Saddam Fedayeen*, and foreign "volunteer" fighters from Syria, Egypt, and other Arab countries. During the day, the division lost an M1A1 Abrams tank to a mobility kill and three more to bad luck, transmission problems and mud, making it one of the worst days of the war for Marine armor. Nevertheless, there was no stopping the advancing Marines, or the tank recovery teams, for that matter, which overcame significant obstacles to get the tanks off the battlefield and back to friendly lines.[309]

By 5 April, the 3d Infantry Division was ready to launch the first of its "Thunder Runs" from the west, its version of the "in-and-out" raids into Baghdad. There was some hard fighting on the way in and out of town. As usual the losses were disproportionate. The Army lost one M1A1 Abrams, but the Iraqi losses were far greater. At the same time, other units from the 3d Division moved west and north to isolate Baghdad.

One of the most amazing things about the "Thunder Runs" is that they, and subsequent forays into Baghdad, were filmed by television crews and broadcast, in real time or near real time, almost as if they were sporting events. But not everyone could watch. The frontline commanders at the regimental level and below did not have much information about what was going on in the "outside" world, the world outside their regiment, perhaps their division; there was a "digital divide" between them and their seniors in the rear, where the amount of information available was simply staggering. The more senior headquarters had a staff with a mind-numbing array of computers, televisions, telephones, and the like. The CFLCC command center at Camp Doha, Kuwait, looked something like mission control in Houston, an amphitheater with enormous screens at the front of the room and rows of desks with computer terminals and telephones. The I Marine Expeditionary Force Main was in the "Bug," the somewhat less comprehensive but still very well connected portable command post. Television sets tuned to news channels were ubiquitous at the higher headquarters, in offices, mess halls, gymnasiums, and berthing spaces. It was almost literally impossible to get away from televised images of the war on the Cable News Network or on the Fox News Channel, whose embedded reporters were popular both with military audiences and with the troops themselves.[310]

On this, the "high" side of the digital divide, there was, arguably, too much information available. It was difficult to know which data stream to enter, how to extract what was relevant, and how to combine streams. Assessment played an important role in managing data. At I MEF, the future operations section was charged with assessing progress toward goals and laying out upcoming decision points. In the vernacular, the calculus was here is where we are, here is where we are trying to go, and here is where we will have to make a decision about the route we need to take in order to get there.[311] At CFLCC, General McKiernan and his chief of staff, General Robert Blackman, had organized the staff by function rather than by traditional Napoleonic section, and created the effects board in order to think of the battle in terms of the effects they wanted to have, rather than the objectives they wanted to seize. Then they instituted the battle update assessment (BUA) in place of the battle update brief (BUB). Virtually every day during the fight, either the commanding general or one of his principal staff officers chaired this virtual staff meeting, which was an attempt to manage the data flow and distill it into a useful form. Run by the CFLCC staff, it was conducted by secure video teleconference and included as participants senior officers (usually flag rank) from all of the major subordinate commands. Eschewing the awkward-sounding acronym "BUB," General McKiernan had opted for the more euphonious "BUA" because he wanted to make the point that the meeting was not about information as much as assessment. He described the battle update assessment as an attempt to lay the foundation for "decision su-

Command in the Information Age

At senior headquarters there has probably never been so much incoming information as there was during the Iraq War, whether it came from official sources or from unofficial sources such as Cable News Network or Fox News. Putting more information into the hands of senior commanders looks, at first glance, like setting the conditions for micromanagement. But generally that did not happen. First, most seniors seemed to exercise restraint the majority of the time. Lieutenant General David D. McKiernan, for one, understood he was there to shape the fight and then leave the details up to subordinate commanders. Second, even if he had wanted to micromanage, the pace of operations generally outstripped the pace of information, especially of good information. General McKiernan commented that since first reports were so often wrong, it was better not to act on them. Finally, even if a higher headquarters had good information and wanted to micromanage, it was often difficult to do so, especially if the subordinate command was moving or was at or below the digital divide. Communications with mobile units or units in built-up areas were just not that good.

Even routine management was difficult. The staffs who tried to stay ahead of the curve and "paper the war" often failed, at least between 20 March and 9 April. Events happened faster than anyone could write old-fashioned operations orders, especially fragmentary orders, before they were overtaken by events. One of the operations officers at the 1st Marine Division, Lieutenant Colonel Paul J. Kennedy, commented that the operations tempo was such that he could not wait for the actual fragmentary orders, but found it possible to work off the preliminary warning orders put out by I MEF.* This was even true at lower levels. At Regimental Combat Team 5, Colonel Joseph F. Dunford found he could not control his command through any kind of systematic planning process after the Opening Gambit.** Interestingly, the special operations community appeared to have an unusually long planning cycle, up to 96 hours, which did not make for happy customers at I MEF. As one senior commander commented, "If they cannot plan and execute any faster, we are going to have a lot of highly trained guys doing flutter kicks in the rear during ops."* Along the same lines, comments were made about the process for generating Coalition Forces Air Component Command sorties being far too slow.

The result was more and more instances of "healthy," straightforward command and control. Imaginative commanders brought "new" products on line to replace the older, barely manageable processes that looked in many ways like a return to the basics. Effects-based planning was one such product. When it worked as intended, General McKiernan's battle update assessment was another. Written or "audible," most operations orders in Operation Iraqi Freedom were about desired effects and reflected the commander's intent. Speaking for division, the operations officer during Operation Iraqi Freedom, Lieutenant Colonel Clarke R. Lethin, made a pitch for communicating that intent:

> How many times have we seen commander's intent developed by the staff, lethargically reviewed by the commander, and then delivered in a briefing without the least bit of emotion? [By contrast,] the division fought by [traditional] commander's intent—a statement of intent that reflected the commander's personality, intuition, sense of purpose, and then [was] delivered to every Marine and sailor in the division.**

Lethin could have been speaking for I MEF as a whole. General James Conway and his subordinates made distinct efforts to communicate their intent, in person, to the extent possible, to every member of their commands. This was one reason for their series of "go-to-war" speeches. They did much the same after the force crossed the line of departure. Although General Conway could have commanded I MEF from Camp Commando in Kuwait, he believed it was necessary to stay forward, to see for himself what was going on, and to meet with his commanders face-to-face. He left the staff work, and the increasingly time-consuming Coalition Forces Land Component Command (CFLCC) battle update assessments, to others and, as his forces advanced deeper into Iraq, moved

*LtCol Paul J. Kennedy intvw, 6Nov03 (MCHC, Quantico, VA).

**Col Joseph F. Dunford intvw, 17May03 (CAT, Studies and Analysis Division, MCCDC).

*Quoted in LtCol Mark M. Tull intvw, 18May03 (CAT, Studies and Analysis Division, MCCDC).

**LtCol Clarke R. Lethin, "1st Marine Division and Operation Iraqi Freedom," *Marine Corps Gazette*, Feb04, pp. 20, 22.

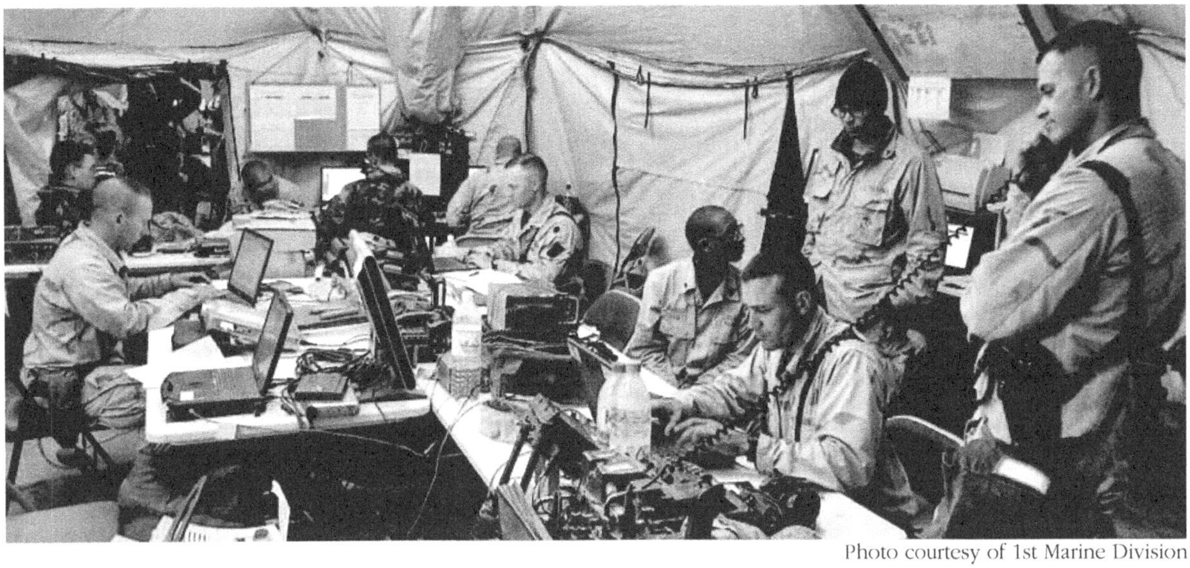

Photo courtesy of 1st Marine Division

The typical forward combat operations center was fully staffed with immediate access to both classified and unclassified digital and analog communications.

the I MEF Forward command post first to Jalibah on 24 March and then to An Numaniyah in early April. Just before the assault on Baghdad, Conway very nearly succeeded in his attempts to visit every unit in 1st and 7th Marines, which were about to engage in the difficult task of subduing the capital city. Almost like a commander in the U.S. Civil War, he wanted to see, and be seen by, his Marines before the battle.* In similar fashion, General James Mattis kept his finger on the pulse of his division during the battle. He separated himself from the "noise" of staff work and, in search of a purer form of warfighting, set off on the battlefield with a tiny retinue and a cell phone. He prosecuted the war as much through personal contact with his commanders, either face-to-face or on the telephone, as through staff channels.

Those who did not meet personally with General Mattis could read his intents messages that were posted, sometimes daily, on the division Web site on the SIPRnet for all with access to this classified network to see. These messages were short, one or two pages long, and written in plain English, outlining the situation as Mattis and his staff saw it and telling readers what the division intended to accomplish. The same was true of some of the slides presented at the battle update assessments and posted on the CFLCC website; the best slides were carefully drafted and approved by the commander himself, containing "balloons" clearly highlighting his intent. Many staff officers began the day by reviewing what was posted on the CFLCC website and those of higher, adjacent and subordinate commands. One staff officer, Colonel Ronald J. Johnson, Brigadier General Richard F. Natonski's operations officer, commented that posting information about Task Force Tarawa on the web enabled I MEF to keep track of the task force and simplified relations between the two commands.*

Transmitting products like the intents messages and the battle update assessment slides, the SIPRnet web made this one of the most democratic wars in the nation's history in the sense that if you had access to a classified computer, as most staff officers and commanders did, you could be empowered by information that in the past was held only by a few officers. The availability of information could make operations that much more efficient. Staffs with access to the same data could plan and operate at the same time without going through time-consuming coordination rituals. Gone were the days when planners had to work in isolation and then pushed products up their stovepipes for their commanders to approve before the commanders passed it to their counterparts. This usually far outweighed the downside, which was that now "an overeager command could monitor every potential contingency and plan for commitments that would never be levied, creating unnecessary confusion and fatigue."**

*LtCol Williard A. Buhl intvw, 4Nov03 (MCHC, Quantico, VA).

*Reynolds, Journal, entry for 1July04.

**Fontenot, et al., *On Point*, p. 11.

periority." It was to enable the commanders to make decisions in the course of their duties after the assessment. He wanted to hear how the major subordinate commands assessed their current and future situations, on the operational level of war, thinking ahead at least 96 hours.[312]*

Both General McKiernan and General Blackman thought the battle update assessment often fell far short of the mark. Some briefers simply recited staggering laundry lists of events as opposed to concise, conceptual summaries of how they saw the battle, and were cut short by McKiernan or Blackman, once with "No, no, no, no—just tell me what I need to know." The various Service cultures were on display during the assessment, with Army commands tending to report in great detail, while the Marines reported more concisely though not necessarily more conceptually, and the British Army seemed to want to report as little as possible. As time went on, the presentations became longer, less conceptual, and more factual, perhaps because of the press of battle. Usually run at 1000 or 1100 local every morning, the assessment could occupy a large chunk of a staff officer's or even a commander's day, since it generated its own set of pre-meetings and post-meetings, and there were complicated slides to prepare on PowerPoint. Some commanders decided to delegate the battle update assessment function to others and seldom appeared on the CFLCC screen.

One thing that distracted staff officers and commanders from assessment was the factual reporting they believed they had to focus on, in part on account of pressure from higher headquarters. Many military reporting procedures had not changed much since World War II, Korea, or Vietnam—the battalion reporting to the regiment to the division, and so forth, a system that worked well enough when the front was relatively static but did not work well when there were multiple, simultaneous lines of operation. It was difficult for units on the move to communicate, especially in built-up areas, because they were busy with other tasks and because the technology often let them down. The I MEF current operations officer, Colonel Dennis Judge, commented that he always knew where his forces were, no thanks to combat reporting, which, he said, "just was not there." Instead he was able to rely on automated systems like the "Blue Force Tracker," which did not require anyone to stop what he or she was doing during combat operations to write and send a report.[313]*

There were others whose job was to report in real time and who had the technology to do so without having to work through a military chain-of-command: the embedded journalists, whose work appeared on the ever-present television screens. The media had been "embedded" (this war's jargon for "assigned") down to the battalion level for the duration of hostilities and had the ability to report directly to their home offices, which in turn were able to rebroadcast their stories in very short order.[314]** What often happened was that higher headquarters would pick up a breaking story on the media and then go back down the chain-of-command to get the official reporting about it, which would painstakingly work its way back up the chain and seemed often to be wrong, at least initially. Some staff officers concluded either that "first reports are always wrong" or that they needed to browbeat their subordinates into doing a better job of reporting. But the problem persisted throughout the campaign. As one CFLCC staff officer quipped, "If you want information bad, you will get bad information."

In early April, as events unfolded around Baghdad at dizzying speed, every commander desperately wanted up-to-the minute information however he could get it, whether from the television newsmen, whose reporting was amazingly vivid and reasonably accurate (on television you could literally hear and see the tanks fire their main guns at Iraqis along the main streets of Baghdad), from the battle update assessments, or even from the official spot and situation reports, which were less timely and far less vivid but a little more accurate.

In the absence of other guidance, the Marines operated in accordance with the next senior commander's intent, which remained simple and consistent: encircle the capital and conduct raids. The I MEF staff officers reported that their plans were to "continue to establish the eastern outer cordon of Baghdad . . . and ...[be prepared to] support further CFLCC missions... [in support of] regime removal."[315] Since there was little by way of amplifying instructions, they had the leeway to exploit opportunities as they saw them. During a conversation on 6 April, Colonel Steven

*The general procedure at CFLCC and other major commands was for operational planning teams, or other groups of specialists drawn from various disciplines, to study problems, propose solutions, and for senior officers to make decisions in small groups.

*The blue force tracker included a transponder that transmitted its location and received the location of all other transponders on the same system, displaying them on a small screen. Since it was not line-of-sight but relied on satellites, it had clear advantages over some "legacy" systems to identify friend and foe.

**By and large, the results of embedding were favorable all around; the reporting had more depth than it otherwise might have, and many journalists seemed more sympathetic to their subjects than they had been in previous conflicts.

©*The Boston Herald*/Kuni Takahashi

Noncommissioned officers of the 3d Battalion, 4th Marines, urge their fellow Marines to cross the damaged Diyala Bridge under fire on the southeast outskirts of Baghdad. The award-winning photograph was taken by The Boston Herald *staff photographer, Kuni Takahashi.*

Hummer of 7th Marines commented that CFLCC had not issued orders telling the field how to conduct raids, and he planned to do it his own way.[316]

On that day, the Army was still planning in-and-out raids. On 7 April, the day of the second "Thunder Run," the 3d Infantry Division reported that it controlled a zone from the international airport to one of Saddam's palaces, with armor at all the main intersections in between, but planned to pull back to the airport. Soon the plan changed; the troopers believed they could hold their ground in downtown Baghdad. This was fine with General McKiernan, who endorsed the idea when General William Wallace told him about the situation. General McKiernan made it plain that it was the corps commander's call; this was a tactical decision, best left to the commander who was in contact with the enemy.[317]

While the Army was raiding into Baghdad, the Marines were working to cross the last major obstacle on their way to Baghdad. That was the Diyala River, located east of the city and flowing roughly north to southeast into the Tigris. It had been briefed as a minor obstacle with numerous fording sites.[318] But this was definitely not the case for the Marines, who found it a challenge just to get close enough to the river to take a look at it. Even with engineering support, crossing did not look easy; the problem was that the riverbanks were either steep, muddy, or both, and the local roads were bad. On the afternoon of 6 April, division reported with some frustration that it had completed a reconnaissance of some 60 kilometers of the Diyala and had "found no viable crossing."[319] It had also encountered stiff resistance in some places.

On 7 April, the frustration continued, preventing forward movements, which may have been a blessing in disguise. General Mattis' comments for that day, and the next, in the daily I MEF sitrep reflected his concern that the logistical situation was "tenuous" and that his command could be nearing a "logistical culminating point," fairly grim words for a very "can-do" commander. He concluded that he was determined to continue "all previous measures to lighten the logistical burden in hopes of building up substantial stocks . . . for future operations," which included conserving artillery ammunition.[320] On 8 April he wrote that the division had only one "day of supply" of food and that it was so low on artillery am-

DVIC DM-SD-04-01787

Engaging the enemy using a P7 amphibious assault vehicle, Marines assigned to Company C, 1st Battalion, 5th Marines, attack an Iraqi sniper position near Baghdad.

munition it had decided to rely more heavily on the air wing than on artillery for counterbattery and preparatory fires.[321] But on 8 April he was able to report that his regiments were all across the Diyala. They had crossed using a variety of means–from seizing bridges, to building bridges, to swimming vehicles across. At two locations, the 8th Engineer Support Battalion played a critical role, putting in assault bridging while under fire, something the Marine Corps had not done for decades. At one RCT 1 crossing site, Iraqi resistance had ebbed as the amphibious tractors entered the water. One of the Iraqi prisoners of war explained why: "When we saw the tanks floating across the river, we knew we could not win against the Americans."[322]

After crossing the Diyala, 7th Marines had gone on to conduct a raid on the Rashid military complex on the outskirts of Baghdad, the ideal intermediate objective, not quite in the city but a good platform for attacks into it when the time was right. At the same time, 5th Marines was attacking first to the north and then to the west to extend the cordon around the city.[323]* The cordon would not be fully complete until 0215Z on 9 April, when the division reported that RCT 5 had cut Highway 2, which ran to the northwest out of Baghdad, and that it had elements of the U.S. Army's V Corps in sight.

By now, most units in the division had shed their chemical suits, the protective overalls in the "woodland" green pattern that contrasted with the desert camouflage uniforms and constricted movement, not to mention how hot they could make the Marine or sailor inside. The general assumption had been that Saddam would use chemical or biological agents outside but not within the Baghdad city limits. The Marines had been on tenterhooks, waiting for the

*According to West and Smith, Col Hummer "forgot" to issue a withdrawal plan for this raid. This is consistent with Hummer interview. The cordon would not be fully complete until 0215Z on 9 April, when the division reported that RCT 5 had cut Highway 2, which ran to the northwest out of Baghdad, and that it had elements of the U.S. Army's V Corps in sight. (West and Smith, *March Up*, pp. 233-234; I MEF sitrep 081800ZApr03 to 091759ZApr03 [Copy in Reynolds Working Papers, MCHC, Quantico, VA]).

Photo courtesy of the 1st Marine Division

Marines from 1st Battalion, 4th Marines, on their first patrol through a crowded community in Baghdad's Saddam City, were greeted by thousands of apprehensive Iraqis who lined the streets.

seemingly inevitable report and enduring the almost eerie absence of bad news. But it never came, and the word was passed that since the Marines were now at the city limits, commanders could make the call on the right protective posture. For most Marines, shedding the unwanted second skin was an incredibly liberating and energizing experience, energy that General Mattis wanted them to use on the enemy in Baghdad.[324]

For their part, the wing and the service support group had continued to do incredible work, pushing supplies to the front, especially through the airhead at Qalat Sikar, which was south of Al Kut, and getting ready to set up shop at another airfield, Salman Pak, on the outskirts of Baghdad. The Marines soon had the food and ammunition they needed, and the cordon around the city was nearly complete, CFLCC's precondition for the next phase. In the words of the command chronology: "By establishing a cordon, the division had closed the door for the enemy's escape, and opened the door to operations inside the city."[325] Now all they needed was permission to proceed beyond their limit of advance.

By 1800Z on 8 April, I MEF was planning to conduct "armed reconnaissance in force into Baghdad" on the next day, and division reported that it was contemplating specific "offensive operations within Baghdad iot [in order to] complete the removal of the Regime." This was quite a change from the day before, when the focus had been on crossing the Diyala.[326] Finally, early on 9 April, the division received the go-ahead from General Conway to attack its targets.[327] A few hours later, the field historian at force headquarters, Major Theodore R. McKeldin III, noted that I MEF had been "cleared hot" by CFLCC for the eastern part of the city.[328] By the time McKeldin penned this note late in the morning, the regimental combat teams had started to move toward the objectives that the division staff had so carefully selected. As one of the embedded reporters, Peter Baker of *The Washington Post*, wrote, they were executing the plan for the day, which "was to keep chipping away at . . . Saddam Hussein's power structure. The Marines would stab into Baghdad and seize a paramilitary base, a secret police headquarters and a presidential palace."[329]

No one was entirely sure what the day would bring. At the morning staff meeting, General Conway commented to his staff that he thought there would be a "big fight" for Saddam's palaces.[330] Division had concluded that the threat from conventional forces had been "nearly eliminated," that there would be little by way of centrally controlled, organized resistance. But there were reports of irregular formations.[331] The regiments were ready to fight if they had to. Soon many of the reports from the front showed quite a different picture. Cable News Network was with 1st Battalion, 7th Marines, as it went into the city, and Fox News was with 3d Light Armored Reconnaissance Battalion to the northeast. Instead of fighters, both battalions encountered crowds of jubilant Iraqis who welcomed their liberation. One of the armored reconnaissance company commanders, Captain Brian B. Smalley, said later that it was "like driving through Paris in 1944."[332]

In his trailer at force headquarters, now at An Numaniyah, General Conway watched the dramatic television footage, some of it showing the poor Shia neighborhood known as Saddam City (and later as Sadr City), which the Marines had originally bypassed. The images impressed him so much that he decided to approve Mattis' request, which had apparently percolated up from the regiments, simply to advance in zone until they encountered opposition. But his optimism was guarded: "Our intent was to constrict the city using various key objectives as lily pads to reduce the regime on the east side. What we are seeing on TV is happening in some Shia neighborhoods. We cannot make the mistake to say that is happening all over Baghdad. We still have a military imperative to conduct operations to reduce . . . regime-related facilities."[333]

Conway discussed the issue with his Army counterparts at a midday video teleconference, and ultimately there was broad agreement on "a plan for a decisive assault on the city" by the Marines and the Army.[334]* This was the end of the various plans for a methodical advance to seize particular objectives, let alone the plan for in-and-out raids. The division finally had a free hand in Baghdad.[335]**

The result was that RCT 7 happily assaulted

Photo courtesy of the 1st Marine Division

Cpl Edward Chin of Company B, 1st Tank Battalion, drapes an American flag over the head of a large statue of Saddam Hussein in the Firdos Square traffic circle. It was quickly replaced with an Iraqi flag on orders from above.

through its assigned area of operations, moving from the southeast to the northwest on the north bank of the Tigris. The list of sites they secured on that day was impressive—the Ministry of Intelligence; the Ministry of Oil; Uday Hussein's offices, which were already burning when the Marines arrived; the Iraqi air force headquarters, which had been destroyed; and the *Fedayeen* headquarters, which was listed as "rubbled." At the end of the day, the division's plans were just as ambitious as its accomplishments: "to have made [the Marine] presence known in all of [our] city zones by morning tomorrow [and] over the next few days . . . allowing the local populace . . . to return to some sense of normal life."[336]

Among the sites in the 7th Marines area of operations was Firdos Square in downtown Baghdad. It was dominated by a six-meter-high statue of Saddam Hussein with his right arm raised in a heroic gesture. The 3d Battalion, 4th Marines, rolled into the square late in the day on 9 April. A crowd quickly gathered. Given the proximity of the local haven for journalists, the Palestine Hotel, there seemed to be as many foreign reporters as Iraqi citizens in the crowd. An Army psychological operations team attached to the Marines arrived and announced over a loudspeaker in Arabic that the Marines had decided the statue should come down. Millions around the world were able to watch the events in real time on Cable News Network and other television networks. A Marine named Corporal Edward Chin, of Company B, 1st Tank Battalion, climbed onto a derrick that extended from his M88 tank retriever. He reached up and placed an American flag over Saddam's face. Some of

*Gen Mattis commented later that the division had "full cooperation" from Gen Conway in his attack on Baghdad. Gen McKiernan also permitted his subordinate commanders a wide range of latitude in tactical decisions.

**The 1st Marine Division command chronology mentions another plan that was supposedly under consideration on 9 April at "higher headquarters." This was for a second operational pause, to give the centralized authority in Baghdad a chance to turn over the city peacefully. The division and I MEF are said to have opposed the plan on the grounds that there was no longer any such authority. It is not clear how division learned of this plan. "Army planners had anticipated that it could take weeks to reduce Baghdad through . . . raids . . . , allowing more time to bring forward military police, civil affairs, and other units required to keep order." (1stMarDiv ComdC, Jun03 [GRC, Quantico, VA]; Atkinson, *In the Company of Soldiers*, pp. 287-288) the Iraqis said, "No, we want an Iraqi flag," and, within one or two minutes, Chin took the flag down and replaced it with an Iraqi flag. Next a stout rope was fitted around the statue, and then to a cable on the tank retriever. When the tank retriever pulled, the statue came down, slowly, as the metal bent and Saddam slipped off the pedestal. The crowd rushed forward, swarming over the fallen statue. One group of Iraqis dragged its head to an unknown but no doubt unpleasant fate.[337]

There are a few arresting images in every war, and the toppling of the Saddam statue was one of them for the Iraq War. It was not exactly what the high command wanted. Generals Tommy Franks and McKiernan, each at his own headquarters, had been watching the scene unfold on television, and when the American flag went over Saddam's face, Franks picked up the telephone to call McKiernan, who did not need to be told why his boss was calling. Even before General Franks could say anything, General McKiernan said, "We are already on it."[338] As commanders from General Franks to General Mattis had told their troops over and over again, this was not a war of American conquest, but of Iraqi liberation. The flag that mattered was not the American flag; it was the Iraqi flag, flown alone, in sites no longer dominated by statues and murals of Saddam Hussein. Nevertheless, the Saddam statue's fall marked a turning point; the end of Phase III seemed to be in sight, even though there would still be some hard fighting.

While the statue had been coming down, Marines a few blocks away had been exchanging fire with regime loyalists, and the next day, 10 April, was no better. On the contrary, 1st Battalion, 5th Marines, spent the better part of the day securing two objectives, Saddam's Almilyah Palace, one of many

A Marine provides cover with his 5.56mm M16A2 assault rifle, with attached M203 40mm grenade launcher, as his squad from Company C, 1st Battalion, 7th Marines, moves through the grounds of one of Saddam Hussein's palaces in Baghdad.

DVIC DM-SD-04-12213

Phase IV in Baghdad

Lieutenant Colonel Christopher C. Conlin, commanding officer of 1st Battalion, 7th Marines, wrote a vivid account of the quick and dramatic shift from Phase III to Phase IV on the streets of Baghdad:

> Within minutes of our seizure of the capital, one of my lieutenants was presented with the rapidly expiring body of an Iraqi who had been pried from a vehicle accident seconds before. He was not a war casualty, but a simple victim of a "routine" incident. The fact that the locals brought him to us as he was in his final death spasms was their unequivocal realization that the normal procedure, going to the hospital or calling the police, was gone. This episode was duplicated throughout our zone with geometrically increasing frequency. Frantic locals ran up to tell us phones were out, doctors reported hospitals being looted, and that the water was off. They were desperate to know where they should dump the trash, could they use cell phones, or was it OK to drive to their father's house in Mosul? They approached us to arrest a strange man with a gun lurking in their neighborhood. We were approached by alleged sheiks who demanded to [see] "his Excellency the General" about their tribe's loyalty to Mr. Bush. . . . On and on the requests came. All being diligently received and somehow answered by our young squad leaders and platoon leaders. We too were in transition. Having overwhelmed organized resistance, we were hot on the trail of Saddam's . . . Fedayeen. . . . Our aggressive small unit leaders were able to hit them in their homes and offices in ever quickening succession. We created hunter-killer teams on the fly, sending our very talented Humint Exploitation Teams . . . PsyOps and Civil Affairs . . . Teams down to the platoon level . . . Companies worked huge urban sectors usually rotating platoons to achieve a 24/7 battle rhythm. The effect was dramatic . . . . Within a week of liberation, the markets were open again, . . . many were going back to work, there was a new police force reorganizing, neighborhoods were organizing . . . Lessons Learned: . . . Press flesh, kiss the babies, and kill the enemy.*

*Col Christopher C. Conlin, "What do you do for an encore?" *Marine Corps Gazette*, Sep04, pp. 74-80.

throughout the capital and the country, and Imam Abu Hanifah Mosque. First they had to drive through a hail of rocket-propelled grenades and "a torrent of heavy machine gun and small arms fire," and then, perhaps the ultimate nightmare, they got lost and had to make a U-turn while under fire. If you think it is bad enough to drive in an unfamiliar city in the Middle East, imagine what it is like to do so with people shooting at you. The Marines were encountering foreign "jihadis," Islamic fundamentalist paramilitary fighters. The mosque was a target because it had been the location of what someone called an "Elvis sighting," Saddam Hussein was said to have entered the mosque, and the Marines were told to secure the perimeter to keep anyone from entering or leaving. The report turned out to be false, but the jihadis still fought with determination. By the end of the day, RCT 5 reported it had lost one Marine killed in action and 42 wounded, along with two tank mobility kills, another unusually bad day for tanks in this war. The commander of RCT 5, Colonel Dunford, remembered 10 April as both the worst and the best day of the war. It seemed like the worst day while his regiment was heavily engaged in downtown Baghdad. His fear was that there would be many more days of heavy fighting. But it turned into the best day when his 1st Battalion reported that the situation was well in hand and that after the fight it was encountering crowds of cheering people, some eight or nine deep, shouting in broken English, "Good, good, Mister!"[339]

The fighting tapered off rapidly, the regime's "last stand" in Baghdad turned out to be so short that it was anticlimactic. There was no sign of Saddam. Some thought (erroneously, as it turned out) he could be lying under the rubble of a restaurant in the Mansour district that had been struck by smart bombs on 8 April. His government had simply evaporated. The civil servants, including "Baghdad Bob," the infamous spokesman who kept predicting on television that Coalition forces were on the brink of annihilation, had simply stopped coming to work. Very soon looters took over the city. If the focus on

DVIC DM-SD-04-08904

Armed with M16A2 assault rifles and 81mm shoulder-launched anti-armor rocket launchers, Marines from Company C, 1st Battalion, 5th Marines, clear the streets of Iraqi resistance block-by-block on their way to securing the Presidential Palace.

9 and 10 April had been on crowds who were literally applauding the collapse of the Saddam regime, it quickly shifted to those who were taking advantage of the situation. The television news broadcast one scene after another of Iraqis, young and old, carrying away anything that was not bolted down from government offices, from official residences, from army bases, even from schools and hospitals. Particularly upsetting for some westerners was the apparent looting of the antiquities museum, a repository of ancient and irreplaceable treasures.

The extent of the looting, and what to do about it, certainly seemed to catch the Coalition by surprise. The basic outlines of OPlan 1003V did not extend far beyond the goals of securing the oil fields, removing the regime, and finding weapons of mass destruction; there had been a great deal of detailed planning, and rehearsals, for Phase III, but the post combat Phase IV had, by comparison, been virtually neglected. If Phase III had been the favorite son, Phase IV was the redheaded stepchild and now he was acting out.[340] When asked about the looting, American officials were low-key. On 9 April, Secretary of Defense Rumsfeld downplayed the problem, as did Generals Conway and Mattis and other senior Coalition commanders. They implied that the Iraqi people had earned the right to profit at the expense of their dictatorial government, and that there was, in any case, only so much the U.S. forces could do, given their limited numbers. General Conway said, "Does it bother me . . . that I see people taking an office chair or the guts of a refrigerator or an air conditioner? Not really, . . . at least [not] at this point. . . . [I]t is got to be the pent-up frustration they've experienced for the better part of their lives."[341] General Mattis offered the uncharacteristically mild observation that the Coalition was "in that kind of never-never land right now . . . we're not in full control [yet], so you're in that transition period."[342]

By the next day, the Coalition forces were shifting their approach and trying to address the problem as best they could with the limited resources at their disposal. Early on the morning of 10 April, a pedestrian suicide bomber blew himself up and injured five men from 5th Marines. The incident caught General Conway's attention. He called it "a sobering reminder

. . . that much work remains to be done." The Marines "must . . . quickly transition to Phase IV Stage B civil-military/humanitarian assistance operations to stabilize the situation and demonstrate our resolve to . . . the newly liberated . . . people" of Iraq. The theme at the I MEF's morning staff meeting was that the Marines needed to switch from a combat role to a stabilizing role. For that reason, Marines were ordered to take steps to control the looting, which included the protection of International Red Cross facilities by 7th Marines.[343] On the same day, 7th Marines took the initiative to establish a civil military operations center in the centrally located Palestine Hotel. The other regiments followed suit in short order, reaching out to local leaders and generally establishing useful contacts in the interests of restoring some semblance of order and basic municipal services. The division's artillery regiment, 11th Marines, played its part. Realizing he was unlikely to need much more artillery support, General Mattis took the unusual step of turning the division civil military operations center over to 11th Marines and assigning the artillerymen their own sector of Baghdad along with the infantry regiments.[344] Late in the day, he was feeling upbeat about the situation: "The city in our zone is . . . [almost] entirely in USMC hands. While it will remain a dangerous place for days to come, the local populace is wholly supportive of our efforts and active patrolling will impose our will. . . . We will calm the situation and . . . build on the goodwill we have experienced to date."[345]

That Baghdad was still a dangerous place was an understatement. The Marines faced a number of daunting tasks from turning the lights back on to explosive ordnance disposal. One problem was, in the words of the division's command chronology, it seemed that "virtually every block" in Baghdad had its own weapons cache. Some of these caches were enormous, "containing every conceivable type of weapon and ammunition, to include tanks, mortars, artillery pieces, and even . . . surface-to-surface rockets."[346] On the march up to Baghdad, Army and Ma-

LCdr Bryan Schumacher, left, and HM Zachary Rowe, of the Regimental Aid Station, 5th Marines, render medical assistance to a wounded Iraqi civilian in the northern Baghdad suburbs.

DVIC DM-SD-04-03523

rine units had, like Task Force Tarawa in An Nasiriyah, found weapons caches throughout the country, often in schools, hospitals, and mosques, to say nothing of the enormous ammunition supply points that covered acres and acres of ground. In the convoluted logic of the Saddam regime, it was only fitting that the capital itself be lavishly supplied and ready for a drawn-out fight. Although Iraqi intentions remained unclear, the discoveries suggested that the regime had been getting ready to back up its threat that the Coalition would face a people in arms, or some kind of guerrilla war, if it invaded Iraq.[347]

Well to the south of Baghdad, Task Force Tarawa had already been conducting Phase IV operations over a vast stretch of country with its approximately 13,000 Marines and sailors. When it expanded its battle space to the north, it became responsible for providing security along the major supply routes, Routes 1 and 7, which meant it had to occupy or neutralize threats in the area. This was in addition to remaining responsible for An Nasiriyah, where Tarawa Marines continued to make notable strides in eliminating enemy resistance and establishing city government. The city of Ad Diwaniyah astride Route 1 posed a typical challenge, which the Marines met in ways that were becoming typical. In early April, various kinds of enemy formations continued to pose a threat to convoys passing through. While cordoning off the city, Tarawa made good use of its Operational Detachment-Alpha, a U.S. Army Special Forces unit that cultivated the population and gathered actionable intelligence. This in turn led to attacks on specific targets with precision munitions. Similarly, in the nearby city of Al Hamsha, information gathered by special operations forces led to a strike against the Ba'ath Party that caused the death of the local Ba'athist leader and was followed by a successful revolt by the local citizenry.

In much the same way, Task Force Tarawa prosecuted attacks against enemy forces in several other cities along Route 7, the other major supply route. These cities were Al Hayy, Qalat Sikar, and Ash Shatrah. Another city that more or less liberated itself was Al Amarah, where elements of the task force had gone to eliminate the *10th Armored Division* from 8 to 12 April in the "Battle of the Icons." In mid-April, after the British expanded their zone to the north and took charge in Amarah, Task Force Tarawa shifted its focus to the cities of An Numiniyah and Al Kut. The task force placed a battalion of infantry in An Numaniyah, where I MEF main had set up an interim command post, and set up its own headquarters on the airfield at Al Kut in hardened aircraft shelters that were mercifully cool, a brief respite from the heat but not from the high tempo of operations.[348]

Chapter 8

No Smell of Saltwater: North to Tikrit, South to Ad Diwaniyah

The expeditionary force's focus had always been on Baghdad. The events of 9 and 10 April signaled the end of the combat operation that most Marines had expected to conduct. Those with longer memories remembered that before the war began, Coalition Forces Land Component Command (CFLCC) had gamed a scenario for securing the oil fields around the city of Kirkuk from the south, instead of the north or the east, and that the I MEF had written a branch plan for Kirkuk.

Some 150 miles north of Baghdad, Kirkuk was the oil capital of the north. The surrounding oil fields were comparable in significance to the southern oil fields and represented an objective far too important to ignore. There had been various plans for the Army to secure the area, especially by introducing the 4th Infantry Division through Turkey. But if that proved impossible, other CFLCC troops might have to advance from the south. After the war, General James Conway remembered being asked by CFLCC if the Marines could take on the mission if necessary, and he answered that it could be done, if necessary. The force was not looking for yet another mission at the end of its supply tether but would take it on if tasked. When asked on 9 April by a journalist about ordering the Marines to advance to the north, General Conway commented: "That would be tough for us. We are a long way from our sea base. As Marines, [if] we do not smell saltwater . . . that makes us a little uneasy. That said, our logisticians, . . . augmented by our [air] wing, have done a magnificent job. . . . To take [it] . . . another couple [of] hundred miles would be extremely difficult. It's difficult right now—that would be extremely difficult. But it might be something that we have to ask our people to do."[349] A few days later, during a staff meeting, Conway made the sensible comment: "We are not putting our hand in the air for that . . . [we] have enough [to do] already."[350]

Coalition Forces Land Component Command's prewar questions meant it was not a complete surprise when the issue came up again and occupied General Conway and his staff on 10 April. The expeditionary force issued a verbal order to division to "be prepared to" proceed to Kirkuk, and to force service support group and wing to be prepared to support the attack. The guidance was to focus on the objective and to avoid being sidetracked by enemy formations along the way.[351]

General James Mattis had already been thinking about the north. Amazingly, he had recently gone on record as being "prepared to continue the attack to Kirkuk . . . and secure the . . . oil field." This was on 8 April, the day when it was worried about running out of ammunition while it was finishing the struggle to get across the Diyala River, just before what would be, for many of its Marines, the fight of a lifetime in Baghdad.[352] Even though division had apparently long had a contingency plan on the shelf for an attack to the north, this was just about as far forward as anyone could, or did, lean in his fighting hole during Operation Iraqi Freedom.[353]

The 1st Marine Division then announced that it could "collapse" its cordon around Baghdad and form a task force, to be known as "Tripoli," after another Marine march across desert sands, that of Lieutenant Presley O'Bannon to Derna in 1805, to conduct the operation. The assistant division commander, Brigadier General John F. Kelly, would command, using the nucleus of the division's forward command post.[354]

Preparations intensified on 11 April. The I Marine Expeditionary Force issued a formal warning order for the continuation of the attack to Kirkuk. Mattis decided on the composition of the force. It would be the division's three light armored reconnaissance battalions: 1st, 2d, and 3d; 5th Battalion, 11th Marines; and Company G, 2d Battalion, 23d Marines. This would amount to the equivalent of a reinforced armored reconnaissance regiment. Logistics would roll with the task force, which would have a dedicated combat service support element; but some resupply would have to come from the wing, which pitched in with plans to establish forward arming and refueling points as the task force moved north and secured suitable pieces of terrain. Cargo helicopters and C-130 Hercules cargo haulers, the workhorse of choice for many missions in Iraq, would deliver supplies to the arming and refueling points. By this point in the campaign most acknowledged that the Marine KC-130s, many of them piloted by reservists, were ideal for the distances and the loads in Iraq; General James Amos was to call them "the queen of the prom" in a postwar

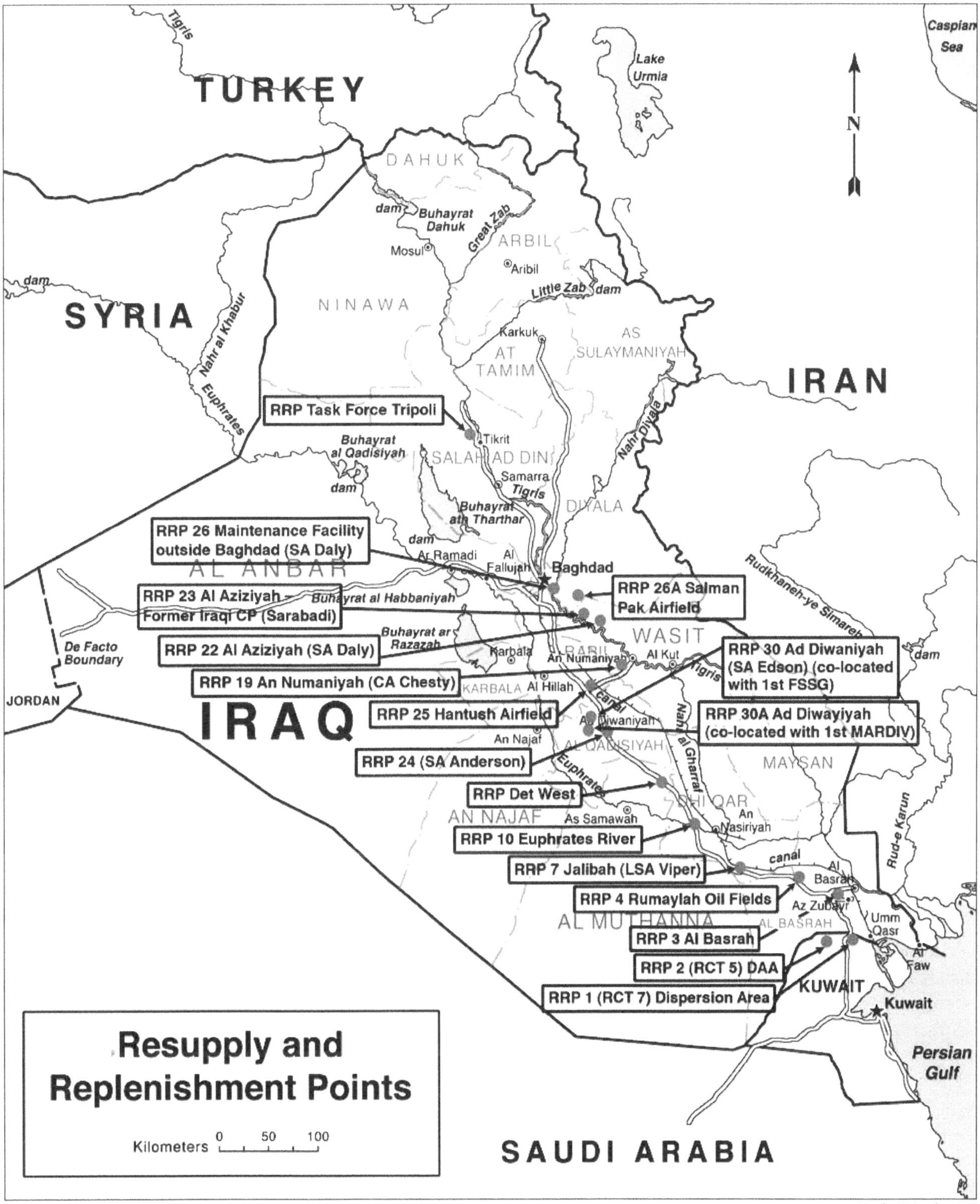

Map by W. Stephen Hill

interview.[355] For its part, the division staff went so far as to analyze the makeup of the oil fields, build objective folders, and recommend schemes of maneuver to General Kelly.

The I MEF plan for Kirkuk was overtaken by events that had been set in motion before 10 April. Kurdish forces, advised by U.S. Army Special Forces teams, seized the oil fields. As of 1800Z on 10 April, Coalition Forces Air Component Command stopped bombing the city of Kirkuk, because organized enemy resist-

The Other Prisoner of War Rescue Operation

Some of the most high-profile stories of the war were about American prisoners of war, especially the rescue of Army Private First Class Jessica Lynch, who had been captured early in the fighting around An Nasiriyah and later became something of a media figure in the United States. On the night of 1-2 April, Task Force Tarawa had facilitated her rescue. But the rescue at Samarra was as memorable, if not as celebrated. It was a victory for the Marine traits of getting out, meeting the Iraqis, and taking calculated risks. On 13 April, human intelligence exploitation team 3, led by First Lieutenant Nathan M. Boaz, entered the city, and an Iraqi policeman approached with information about U.S. prisoners. To pinpoint the location without compromising the source, the exploitation team's chief, Staff Sergeant Randy Meyer, gave the policeman a global positioning system receiver and told him how to use it to record the location of the Americans, which the policeman did. Even though he was under pressure to continue on to Tikrit, 3d Light Armored Reconnaissance Battalion commander, Lieutenant Colonel Herman S. Clardy III, decided on his own that the information was reliable enough to act on, and that trying to rescue the prisoners was worth the time. Armored reconnaissance scouts and exploitation team members proceeded to storm the house pinpointed by the informer and, without firing a shot, freed the seven grateful soldiers—two pilots and five members of Jessica Lynch's unit, who had also been captured in An Nasiriyah on 22 March. One of the pilots later described the rescue on television in vivid terms, cheerfully recounting that he had never before been so glad to see a group of men who looked like hollow-eyed killers coming to get him. The news of their release was immediately briefed throughout the theater, and the prisoners were whisked to the rear.*

DVIC DM-SD-04-07386

LCpl Curney Russell and Cpl Christopher Castro from 3d Light Armored Reconnaissance Battalion provide a steady arm for rescued prisoner of war, U.S. Army Spec Shoshana Johnson of the 507th Maintenance Battalion, as she and other repatriated prisoners are led from a Marine KC-130 aircraft at Kuwait City.

*In addition to the division command chronology, see LtCol Herman S. Clardy III, email to author, 28Apr04 (Reynolds Working Papers, MCHC, Quantico, VA).

ance was crumbling. On 11 April, the city was in Kurdish hands. By late afternoon on the same day, CFLCC changed the Marines' target to Tikrit, Saddam Hussein's hometown, which was only about 100 miles north of Baghdad. General Conway commented to his staff by way of explanation that the Marines "were the best force available now to do this."[356]

Tikrit was assessed by various intelligence elements as a logical place for the regime to make a last stand; the division framed the operation as an attack to deny the enemy safe haven and to prevent the escape of high-value targets. The enemy forces that the division expected to encounter were one or two brigades of the *Special Republican Guard* and a few hundred paramilitary fighters. The general assumption was that the paramilitaries would fight but that the uniformed soldiers would not.

The change may have been frustrating to some forces and division planners, but overall it was a positive development, since it was relatively easy to substitute Tikrit for Kirkuk, and Tikrit was closer to Baghdad; the supply tether would not have to stretch as far. In General Mattis' words: "The division is stretched, but not to the breaking point, with stabilization ops in Baghdad [and] TF Tripoli focused on Tikrit. . . . We are green on logistics, but will require a new . . . estimate of supportability if ordered to the Kirkuk or Mosul area."[357]

What General Conway called "the final attack" officially began on the afternoon of 12 April, at 1255Z, to be precise and proceeded rapidly up-country. It was, in a sense, a quintessential Operation Iraqi Freedom operation. It was short-fused and had a commander who stressed the need for speed. He

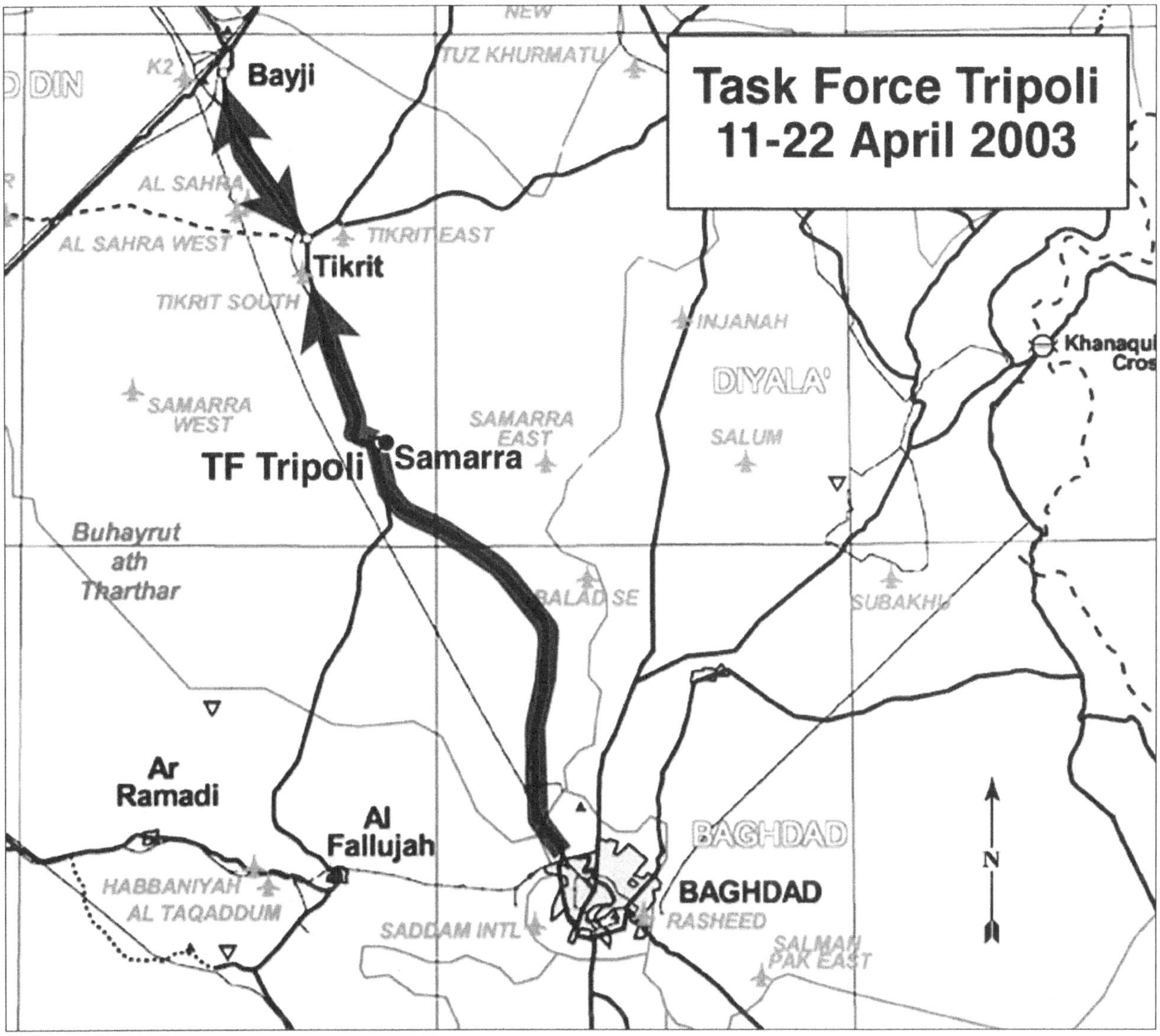

task-organized in a matter of hours, making good use of two weapons systems that also seemed to have been made for a place like Iraq, eight-wheeled light armored vehicles and Cobra helicopter gunships. "It was," the command chronology recalled, "comforting [to have] the Cobra escort. They were masters at rooting around the enemy's potential hiding spots and building our situational awareness around every corner. . . . We gained a deep respect for the Cobra pilots during the war."[358] The wing's airborne forward air controller complemented the work of the Cobras and kept in touch with a section of Harrier jets that were dedicated to the task force. When this piece of division-wing coordination was reported to I MEF, the word was that if anyone opened fire on the Marines, he would not live to regret it.[359]

After leaving its assembly area, Task Force Tripoli passed first through friendly territory, with Iraqis lining the roads to wave and shouting greetings like "Good, good!" In Samarra, a city between Baghdad and Tikrit, there was a happy event when the 3d Light Armored Reconnaissance Battalion found and freed seven American prisoners of war on 13 April. Then, closer to Tikrit, the locals became less friendly and in some cases downright hostile. There were even more murals, portraits, and statues of Saddam here than in other parts of the country, where they were certainly plentiful enough, but the Marines noticed only a few had been defaced. Both on 13 and 14 April, there was "minimal contact" with enemy formations, occasional firefights, some of them intense, on the way to the objective.

"Minimal contact" is the kind of phrase that senior commands put in situation reports. To the Marine on the ground who is risking his life during one of the few firefights of the day, it certainly does not seem "minimal," and the division's command chronology contains vivid descriptions of fights with *Fedayeen* in and around Tikrit. Even by the Iraqi standard, which seemed to call for every school and mosque to be filled with munitions, there was a great deal of military hardware in the area. The Marines of 1st Light Armored Reconnaissance Battalion, for example, passed through a "large area [south of Tikrit] . . . with multiple revetments that contained enemy vehicles" and then called in air strikes on the vehicles before proceeding on to a crossroads where they took a number of rocket propelled grenade rounds, one of which struck the tailpipe of a light armored vehicle. The Marines returned fire, destroying two trucks and killing five of the enemy. Once inside the city, the Marines found massive amounts of ordnance, including a compound with 30 warehouses of weapons and a cache at a hospital containing literally thousands of AK-47 rifles. The division concluded that "the amount of ordnance found in our sector is overwhelming and will require months of dedicated lift to remove it all."[360]

By 15 April, fighting had tapered off, and Task Force Tripoli found itself the de facto source of power in Tikrit. The question was how to use that power to address the bewildering array of problems that the city faced. The existing power structure, the tribal sheikhs, approached the task force, outlined the situation as they saw it, and wanted to establish a leadership council. Brigadier General John Kelly was reluctant to deal with the sheikhs, who were hardly a democratic body and had done business with Saddam Hussein. In the end, he decided that the sheikhs represented as good an interim local government as any and met with them on 15 April to lay out his view of the situation. He repeated a number of basic themes: the Marines would take responsibility for the security of the area; the locals would help the Marines hunt down any remaining foreign fighters or Saddam loyalists; and, finally, the sheikhs would bear much of the responsibility for restoring power, water, food, and medical services. All of this was easier said than done; the 15 April meeting was simply the beginning of a process that entailed numerous meetings between the sheikhs on the one hand and General Kelly on the other, as men conditioned by long-simmering feuds and unaccustomed to democracy took their first tentative steps away from a dictatorship that had dominated their lives for 30 years. A by-product of the good relations between General Kelly and the sheikhs was their assistance in the peaceful "liberation" of the city of Bayji about 25 miles north of Tikrit.

MSgt David Livingston from Marine Wing Support Squadron 271 supervises a KC-130 Hercules offload of fuel at a forward arming and refueling point.
DVIC DM-SD-05-05193

Meanwhile, out on the streets of Tikrit, the Marines dealt with such challenges as a panicky crowd of citizens who wanted to cross over a bridge that had been weakened during the war; a group of vigilantes who established their own checkpoints outside the city, apparently to harass people they did not like; and marauding Kurds, the ethnic minority who had suffered at the hands of Saddam's regime and were not inclined to treat the citizens of Tikrit gently. The Marines' aggressive patrolling, typically by dismounted infantry without flak vests and in soft covers, along with their commonsense solutions to problems, restored more than a small semblance of order to the city within a few days.

Between 19 and 21 April, Task Force Tripoli conducted a relief in place with the Army's 4th Infantry Division and prepared to rejoin the main body of 1st Marine Division. The 4th Infantry Division had spent the war en route first to Turkey, where it had been unable to land, and then to Iraq through Kuwait. Although major combat operations had ended, this division characterized its move north to the zone it would occupy as an "attack," which led to some unfortunate incidents. On 19 April, the Marine division complained about U.S. Army Apache helicopters entering Marine battle space "without permission or clearance to engage," and then firing at abandoned enemy armor located between the 2d and 3d Light Armored Reconnaissance Battalions. The 4th Infantry Division also seemed intent on treating the Tikritis as hostile, which distressed the Marines, who had gone to some lengths to establish good relations with the locals. General Mattis himself noticed that like the 3d Infantry Division, the 4th had a lot of armored vehicles and only a limited number of dismounts and went on record with a prediction that "the lack of Army dismounts is creating a void in personal contact and public perception. . . . Our forces need to project confidence in the security environment we have created. That is best exemplified in light, mobile forces in contact with the local citizenry sans helmets and flak jackets and without armored vehicles. If we cannot engender friendship and confidence in the local security environment, we cannot set the conditions for good order integral to a return to civil control."[361]

General Mattis' comment applied not only to Tikrit but also to Baghdad. While Task Force Tripoli was securing Tikrit and preparing for a relief in place with 4th Infantry Division, three regiments of the division had been consolidating their gains in Baghdad and getting ready to turn their zones over to 3d Infantry Division in accordance with the plan for the occupation of Iraq. The civil-military operations center run by the artillerymen of the 11th Marines, continued to find stopgap solutions for the many problems they faced; with remarkably little assistance or guidance from anyone, it had established cells for police, fire, electricity, water, and medical care, and the instant experts worked their issues as best they could. The operations center appointed an interim police chief and opened an Iraqi police academy. It facilitated and coordinated the work of nongovernmental organizations like CARE, the Red Cross, the Red Crescent, and the World Food Program. Meanwhile, out on the streets, the regimental combat teams maintained checkpoints to control movement in their sectors, which they moved often to frustrate any attempts to target them. The division took some additional steps such as delivering public safety messages, in the hope of controlling the rampant looting and widespread crime that had broken out when the regime collapsed and left a power vacuum in Baghdad. Since this was something no one had really planned for except in the most general terms, there had been an ad hoc quality about the guidance from CFLCC and I MEF about looting and crime that seemed to boil down to "stop it if you can" and left it up to the units in contact to come up with workable solutions.[362]

Even though there was only so much the Marines could do in the few days between the collapse of the regime and their scheduled departure from Baghdad, they appear to have made some progress. One Marine, Major Jason L. Morris, of 3d Battalion, 5th Marines, observed that the eastern half of Baghdad seemed to have returned to something like normalcy within 10 days of the fall of the Saddam statue on Firdos Square. Morris attributed this to the division's approach, which he characterized as relatively low impact. Even in the days immediately after 9 April, Marines allowed the local citizens to go about their business so long as they did not interfere with Marine business. There was certainly an active Marine presence in the streets, especially by dismounted in-

DVIC DM-SD-04-16619

Marines assigned to the 1st Light Armored Reconnaissance Battalion encamp below one of Saddam Hussein's palaces in Tikrit, the Iraqi president's hometown.

fantrymen, but he did not think they gave the city the feel of being under martial law. He supported General Mattis' argument that infantrymen are better suited for patrolling a city than armored vehicles.[363]

As impressive as its beginnings were, no one will ever know how effective the division's occupation of Baghdad would have been in the long run. By 15 April, if not before, the division staff had begun to prepare for a turnover to the 3d Infantry Division, with the assistance of the Army's 358th Civil Affairs Brigade's Deliberate Assessment Team. Detailed staff work and command interaction followed at various levels, more or less spontaneously, "without written orders, instructions, or doctrine of any kind," to quote the division's command chronology.[364] The actual relief in place in Baghdad occurred between 18 and 20 April, another eventful day. While elements of the division moved south to Ad Diwaniyah, where it relieved the Army's 82d Airborne Division and established a new headquarters, I MEF convened a combined Phase IV and retrograde-reconstitution-regeneration conference at its dusty field headquarters in An Numaniyah.

Phase III, the war from the line of departure in Kuwait to the enemy center of gravity in Baghdad, had gone by in the blink of an eye, a remarkable performance by any standard. But there had been a cost. For the Marines, the casualties were on the order of 75 killed in action and 300 wounded. Even if statistics did not lessen the heartbreak of loss for each man who died, the numbers were far smaller than had been feared, which was something to be thankful for.[365] The Marines honored their dead and wounded and prepared for the next phase of the operation.

Chapter 9

The Other War in the North: Marines in Kurdistan and Mosul

Task Force Tripoli was not the only Marine command that operated north of Baghdad, which included great stretches of territory that fell under Joint Special Operations Task Force North, commanded by Colonel Charles T. Cleveland, USA. Cleveland, who had apparently been in Iraqi Kurdistan for months, reported not to Coalition Forces Land Component Command (CFLCC) but to Coalition Forces Special Operations Command, another CentCom dependency. Using various kinds of paramilitary forces, he ran the war in the northeast corner of Iraq, responsible for its defense from the Iraqi Army and for preparations to secure the cities of Mosul and Kirkuk and their surroundings. He had, in a sense, taken on the mission that had been assigned to the 4th Infantry Division, to put pressure on Baghdad from the north, or at least to manage the situation in the north in order to prevent any surprises that might undermine the offensive in the south. To do this he had at his disposal a U.S. Special Forces Group, made up of three Special Forces battalions, and Kurdish formations as well as Coalition airpower in general support. There was a U.S. Air Force contingent with some air and ground assets complementing the Special Forces, and Colonel Cleveland could also request sorties from Coalition Forces Air Component Command. As for the Kurdish fighters, there were some 70,000 of them, grouped into two militias identified with the two main Kurdish political parties, the Patriotic Union of Kurdistan and the Kurdish Democratic Party, which were now cooperating against the common enemy but had in the past sometimes turned on each other, exhibiting some of the less attractive aspects of a warlike culture. The Special Forces had been working with these fighters for at least a few months.[366]

There was another group working in the north that did not come under Cleveland but cooperated with his command. It was the handful of Marines known as the Military Coordination and Liaison Command, led by Major General Henry P. Osman, the commanding general of II Marine Expeditionary Force from Camp Lejeune, North Carolina, another one of the low-key but very effective generals who left their mark on Operation Iraqi Freedom.

The Military Coordination and Liaison Command had a complicated history. In the fall of 2002, General Osman had watched Brigadier General Richard Natonski and 2d Marine Expeditionary Brigade's preparations to join I MEF in Kuwait for the attack from the south. This was neither unexpected nor a particular hardship to II MEF; since the end of the Cold War, and the shift in focus away from the North Atlantic, II MEF had served largely as a force provider for other theaters. The 2d Marine Expeditionary Brigade was by design a detachable module of II MEF. General Natonski himself was the deputy expeditionary force commander, as well as the commander of the brigade. Similarly, many of his officers were dual-hatted, with both force and brigade functions. The deployment of the brigade meant the force staff had only a small residual capability, but it was still a capability that could be put to good use in the

Photo courtesy of Field History Branch

MajGen Henry P. Osman, a graduate of Old Dominion University, was commissioned in 1969. Over the ensuing years, he commanded a rifle platoon and company in Vietnam, a 1st Marine Division infantry battalion, and served in numerous staff assignments before assuming command of II Marine Expeditionary Force in August 2002.

CentCom theater. General Osman believed that II MEF could field a small joint task force and told Headquarters Marine Corps as much.[367] What happened next was a dialogue between II MEF and Headquarters Marine Corps over the potential uses of II MEF's staff, animated at least in part by the Commandant of the Marine Corps, General James L. Jones' continuing interest in northern Iraq, especially the northeast corner of the country with its large Kurdish minority.

In 1991, in the wake of Operation Desert Storm, there had been a humanitarian crisis in northern Iraq following Saddam Hussein's brutal suppression of a Kurdish revolt. After a delay, President George H.W. Bush had directed the Commander-in-Chief, U.S. European Command to conduct an operation to provide humanitarian assistance to the tens of thousands of Kurds who, in the aftermath of the revolt, had become refugees in their own homeland. The result had been Operation Provide Comfort, a multinational effort that had lasted for about four months and created the basis for de facto Kurdish autonomy for the next decade. Iraqi troops stayed out of Kurdish territory, part of the northern no-fly zone enforced by Operation Northern Watch. That said, in 2002 there were up to 12 Iraqi divisions north of Baghdad, at various stages of readiness, grouped in *I Corps* and *V Corps*. A number of them were assessed as capable of advancing into Kurdistan. Both the Iraqis and the Kurds actively manned and patrolled the "Green Line" that divided their spheres of control.[368]

The Commandant had been the 24th MEU (SOC) commander during Provide Comfort, and he had not forgotten the experience. Now, in late 2002, as he prepared to step down as Commandant and take up the responsibilities of Supreme Allied Commander, Europe, he wanted to do what he could to support CentCom in the coming war against Iraq. One of his agenda items was to explore the need for another such mission with Marine participation.[369] Would the CFLCC attack from the south lead to another humanitarian crisis in the north? If so, it would involve Turkey, a NATO ally. Southern Turkey was the obvious location to stage the mission. This was therefore a matter that the Supreme Allied Commander, like his predecessor 13 years earlier, needed to consider. The border between Iraq and Turkey was also the border between CentCom and European Command; anything that happened in northern Iraq would affect both commands. General Osman and the II MEF staff were natural candidates for a military-political mission to the region, and after he had more or less volunteered his services, Osman discussed the contingency with General Jones in November and December 2002. He also met twice with General Anthony Zinni, another veteran of Operation Provide Comfort, who, while at Central Command, had spent much of his time traveling around the region practicing military-political diplomacy.[370] By late January 2003, after 2d Marine Expeditionary Brigade had left Camp Lejeune, General Osman was in touch with the senior Marine in U.S. European Command, Marine Forces Europe commander, Lieutenant General Martin R. Berndt, about standing up a joint task force to coordinate Phase IV, that is, post combat, operations in northern Iraq. To consider, first, how to staff the joint task force and, second, how it should operate, II MEF had created an operational planning team.[371]

On 17 February 2003, General Osman and a few of his principal staff officers flew to the European Command headquarters at Stuttgart, Germany, to discuss the potential operation with military planners at Marine Forces Europe, which is part of European Command. Next they visited Ankara, Turkey, for meetings with subject matter experts at the American embassy and the Turkish General Staff. To put it mildly, the Turks had no enthusiasm for a mission to help the Kurds, especially one operating out of southern Turkey, and did not offer much by way of help or encouragement. Since they wanted to keep their own Kurdish minority in check, the last thing the Turks wanted was for the Iraqi Kurds to assert themselves. The Turks and the Kurds had a long tradition of enmity, and the United States had to strike a delicate balance between using the Kurds against the Iraqis and not upsetting Turkish sensibilities. This was one of the reasons why the United States could not use the Kurdish militias as proxies in the quite the same way it had used Afghan militias in late 2001 and early 2002. Throughout the operation, as General Osman and his officers encountered strong, even visceral, anti-Kurdish sentiments among the Turkish officers they met, they made it part of their mission to do what they could to keep such sentiments in check. It did not help that General Osman had a typically Turkish surname but was not of Turkish descent, leading his Turkish contacts first to imagine that they were speaking to a compatriot, and then to disappointment when they were disabused of that notion.

General Osman's small group went on to Camp Doha, Kuwait, to meet with General David McKiernan and his staff, whom they found to be, understandably, preoccupied with other matters and not able to devote much attention to northern Iraq. The group's last stop was CentCom headquarters at Doha,

Qattar, where there were more discouraging meetings. A down-to-earth son of Georgia who called it like he saw it, Colonel Robert L. Hayes III, the I MEF operations officer, remembered thinking that no one really seemed to want the mission to happen. He himself was beginning to question its utility, especially since his mission analysis showed that the conditions that had called forth Operation Provide Comfort in 1991 simply did not exist in 2003. There did not appear to be an impending humanitarian crisis; on the contrary, this time the Kurds appeared to be well prepared for any such eventuality. The joke among members of the staff was that this task force without a mission should be known as "JTF JDK," for "just do not know." Someone even went to the length of ordering coffee cups with an official-looking "JTF JDK" logo.

Virtually at the last minute, during a meeting with General John P. Abizaid, USA, the deputy commander of CentCom forward and another veteran of Operation Provide Comfort, the mission was saved. General Abizaid analyzed the current situation from another angle and came to the conclusion that the presence of the right American flag officer would definitely be value added. That officer's mission would be to preserve the status quo during combat operations, in effect, to help secure the northern flank, and not just to prepare for humanitarian assistance after the war. The officer would work to keep the Kurds (and the other minorities) in check, which in turn would keep the Turks from intervening "to protect their equities," a distinct possibility if the Kurds seized Kirkuk, the medium-sized city that was the key to the northern oil fields, and Mosul, its larger neighbor to the north, with a population of more than 1.5 million that many senior Ba'athists called home. Even without the Turkish factor, there were so many different groups with ancient conflicts against one another that the region was a powder keg. The Turkish threat to intervene if the Kurds "got out of hand" was the match that could have set it aflame. With this vision, General Abizaid played the leading role in converting the European Command initiative into a CentCom mission and became the mission manager to whom Osman reported, on an almost daily basis. General Abizaid told General Osman that the best place for the Military Coordination and Liaison Command was in northern Iraq, not southern Turkey.[372]

Even though Kurdish Iraq was more or less autonomous and enjoyed the protection of its own forces, this was still enemy territory. Getting ready to operate in Saddam's backyard during the war made matters more interesting and it called for a lot of creative preparations. Leaving roughly 30 members of his staff in Stuttgart, General Osman decided to enter Iraq with only a handful of Marines, a total of six. Through a portable communications suite, they would "reach back" for support to Stuttgart from members of the II MEF staff who had deployed to European Command, and rely largely on the Kurds and U.S. Special Forces for situational awareness, security, and other forms of support. They acquired a mix of civilian sports utility vehicles and one pickup truck, a Ford F-350, through a mix of open purchase and barter with the U.S. Air Force. They planned to travel in uniform, with a mix of light weapons. Everyone on the tiny staff had more than one job. Colonel Keith A. Lawless, for example, was not only the intelligence officer (his normal job), but also the political adviser and a bodyguard armed with an M4 assault rifle. Colonel Lawless' colorful operations chief, Master Gunnery Sergeant Richard C. McPherson, doubled as an interpreter in addition to his normal duties, and as driver and bodyguard. Born in Jamaica to Scots-Lebanese parents, McPherson had served a number of tours in the Middle East and was an accomplished Arabist.[373]

Just getting the approvals to travel through Turkey, and then to cross the Turkish border into Iraq, was no simple matter—requiring at one point the personal intervention of General Jones, the Supreme Allied Commander, Europe, with the Turkish high command. Finally, on 23 March the small task force made it out of Turkey and found itself very much alone on the Iraqi side of the deserted border crossing, until two heavily armed U.S. Army Special Forces noncommissioned officers appeared, seemingly out of nowhere, to serve as escorts and advisers. They made no secret of their initial lack of enthusiasm for the newcomers who were taking them away from their primary duties of training the Kurds and helping them to kill Iraqis. But, like others in Joint Special Operations Task Force North, they soon discovered that General Osman's group was low-maintenance and that its work could complement that of the Special Forces.[374]

Within two hours of crossing the border, Osman's work started in earnest. At a town called Dahok near the Green Line, he stopped to meet General Babekir al-Zibari, the senior KDP military leader in the region, whom General Osman came to appreciate as "a great warrior and a great . . . patriot."[375] Babekir's troops, and those of other Kurdish organizations, were known generically as the "Peshmerga," literally, "those who face death." The average Peshmerga fighter looked like a cross between a mountain man and a horseman, but there was little doubt about his motivation or his fighting ability. Peshmerga groups ranged from loosely organized militia to semiprofessional paramilitaries, with about 45,000 loyal to the Kurdish Democratic Party and some 20,000 answering to the Patriotic Union of Kurdistan.

Over the course of a meeting that lasted many hours and included an elaborate lunch, General Babekir talked about his appreciation for what the United States was doing in the region, and especially for the work of the U.S. Special Forces, with whom he had a strong ongoing relationship. He said he was eager to do more, which was his way of saying he wanted to cross the Green Line and fight the Iraqis. In what was to become a familiar refrain, General Osman counseled patience, explaining the instability that Kurdish action could unleash, given Turkish sensibilities. General Babekir took this on board and asked if General Osman could do anything about the Iraqi artillery on the other side of the Green Line that could range Dahok. Osman made no specific promises but said he would see what he could do. By chance, a few hours later a heavy Coalition air raid was made on the Iraqi lines, and General Osman's penny stock with the Kurds increased in value a hundred fold. The result was, in the words of the mission's communications officer, Lieutenant Colonel James E. Bacchus, "effervescence" from the Kurds, who felt they had been abandoned by the Americans in the past. Perhaps this time it would be different.[376]

General Osman and his party went on to the town of Salaheddin, where they stayed in a guesthouse belonging to the Kurdish Democratic Party, the somewhat more westernized, and larger, of the two Kurdish political parties. The next day, 24 March, General Osman held a press conference, there being a large contingent of media in Salaheddin, during which he explained the Military Coordination and Liaison Command's mission in general terms, outlining such functions as "the deconfliction of military activities," synchronizing humanitarian assistance, and "the maintenance of stability."[377] Reports of the conference described the liaison command as a "special military command to protect northern Iraq and satisfy Turkish security concerns along the Turkey-Iraq border," and especially to "dissuade Turkey from sending troops to northern Iraq."[378] The presence of a two-star Marine general had clearly caught the attention of the players on the field and demonstrated that the United States wanted to make sure the game was played by its rules. It also contributed to the "information operations" against the Iraqis, who had to be asking themselves what a senior American officer was doing in Kurdistan, was he paving the way for a larger force? General Osman did nothing to discourage this speculation, especially when he reached out to the Iraqi generals on the Green Line, passing a very direct written message from General Tommy R. Franks: "Decide now. I, General Franks, will help you destroy Saddam's regime, or I will destroy you."[379]

The next few days saw a round of seemingly endless meetings with a bewildering array of local groups, from the routine, such as coordination with Joint Special Operations Task Force North, to the conventional, such as meetings with the Kurdish Democratic Party and the Patriotic Union of Kurdistan, to the exotic, such as with the Iraqi Turkomen Front, representing the Turkish minority in the region, and the Supreme Council of Iraqi Revolutionaries, an anti-Saddam group supported by Iran. Most of the time General Osman drove long distances to meet the local politicians and functionaries, choosing to meet them on their own turf rather than summoning them to his headquarters; he wanted to avoid the appearance of seeming like an imperial envoy. This entailed some risk of attack on the road from Iraqi operatives, who were said to have targeted the general for assassination. Although sometimes accompanied by a small Kurdish security detail, along with one or two U.S. Army Special Forces soldiers, force protection for the small group was mostly a matter of not making its destination known until the last minute and then traveling at breakneck speed on bad roads. In the end, the general told his drivers to slow down; he said he would rather die from a sucking chest wound than from the seemingly inevitable car crash. Once they arrived at a home or office, there were elaborate rituals of courtesy to follow, which typically meant drinking a lot of tea and eating large quantities of unusual food. Some of the food was easily identifiable, like roast lamb and rice, two local staples. But there were also unusual foods, like one that a member of the mission could only describe in his notes as a "strange veggie." This routine, and General Osman's personal style, which was characterized by patience, politeness, and persistence, enabled him to build relationships with virtually all of these groups. Although Osman did not supplant Colonel Cleveland, he became something like the senior adviser to call in order to discuss a plan, or the honest broker who could keep the peace between warring factions.

The value of these relationships emerged as the Iraqi regime collapsed at the end of the first week in

April, and the Iraqi soldiers in the north either surrendered or simply changed into civilian clothes and went home, like thousands of their comrades throughout the country. One result of the collapse was a power vacuum in Kirkuk and Mosul that some Kurds wanted to fill. There had been substantial Kurdish minorities in both of these Iraqi cities, and many Kurds had lost property when Saddam Hussein had "Arabized" the region. Understandably, the Kurds wanted to protect and assert their interests in both cities, while the Turks wanted to make sure the Kurds did not turn them into centers of Kurdish power. Apart from its interest in removing vestiges of the former regime, especially in the "loyalist" city of Mosul, and stabilizing the region, the United States was as dedicated to protecting the oil fields around Kirkuk as it was to protecting the southern oil fields around Basrah. With its blue-collar outlook, Kirkuk reminded General Osman of a frontier oil town, while Mosul seemed to be more developed and sophisticated, even attractive in parts.

The Kurds became increasingly anxious as rioting and looting spread through Mosul, where there were few Coalition troops in evidence. In Kirkuk, the Kurdish Democratic Party and its Peshmerga took the initiative to fill the vacuum. Its leader, Jalal Talibani, literally moved into city hall. His Patriotic Union rival, Massoud Barzani, immediately complained to General Osman that Talibani had broken the rules and expanded his sphere of influence at Barzani's expense. Barzani claimed to be losing ground because he had listened to General Osman's pleas for restraint. Osman's response was to drive to Kirkuk on 12 April, through a kind of no-man's land, and find Talibani, whose reaction on seeing the general was to try to embrace him in order to celebrate the day the Kurds had returned to Kirkuk. Within seconds Talibani saw that the general did not share his mood. The two stepped into the former mayor's office, where General Osman told Talibani in no uncertain terms why he and his forces had to leave Kirkuk. If they stayed, the result could be Turkish intervention, and a war within a war. This was no idle threat; there were Turkish liaison officers and a few Turkish special forces soldiers on the ground pursuing their own, crosscutting interests. Talibani, who spoke good English and wore western-style clothes, reluctantly came around to the general's point of view and agreed to withdraw himself and his troops from Kirkuk. Even General Osman's staff was impressed with this result, remarking, "Gee, Sir, you threw Talibani out of town."[380]

The townspeople of Kirkuk had seemed to be favorably disposed to the Coalition and eager to return to a peacetime routine. At first they had not known what to make of the unusual group of Marines driving into town in black sport utility vehicles. But when they had realized they were looking at Americans, they had started to applaud. Mosul, by contrast, felt hostile, not unlike Tikrit. The locals appeared glum, and there were Iraqi flags flying everywhere, even from mosques. It did not help that Peshmerga elements were active in the city, which inflamed tensions between Iraqis and Kurds. While the city was quieter on 13 April than it had been the day before, General Osman agreed with task force officers that pockets of unrest were still apparent and that the United States needed to increase its military footprint as quickly as possible.[381]

Photo courtesy of Field History Branch

Secretary of Defense Donald H. Rumsfeld, center, talks with Massoud Barzani, left, president of the Kurdish Democratic Party, in Salahuddin, Iraq, during meetings with Kurdish party leaders.

As if on cue, the 26th MEU (SOC) appeared over the horizon. It was the only readily available, uncommitted force in the area and was still in the Mediterranean, under the cognizance of Sixth Fleet, which meant it was ultimately under the command of General Jones. It had been natural for him, a Marine with an interest in northern Iraq, to discuss and plan its employment with CentCom. General Jones had always been a keen proponent of putting a Marine air-ground task force into northern Iraq, and a Marine expeditionary unit fit the bill.[382] Commanded by Colonel Andrew P. Frick, the 26th MEU was another unit that had had an incredible existence, so different from most peacetime expeditionary units that simply went from one landing exercise to another until their time was up. The 26th MEU had flown from expeditionary shipping into Afghanistan the year before in what was viewed at the time as validation of the Marine doctrine and was now about to launch a new long-distance operation over semihostile territory.[383]

Part of the Marine expeditionary unit soon flew by Lockheed KC-130 Hercules aircraft from Naval Air Station, Souda Bay, on the island of Crete, to northern Iraq, a distance of some 975 nautical miles, while its air combat element, Marine Medium Helicopter Squadron 264 (HMM-264), self-deployed from the USS *Iwo Jima* (LHD 7) in the eastern Mediterranean to fly nearly 500 nautical miles into Iraq, possibly a new record. The unit historian, Captain Arnaldo L. Colon, noted that the "HMM-264 self deployment into Iraq was the farthest a MEU (SOC) ACE [aviation combat element] had ever flown. Prior to that flight, HMM-365 had flown the farthest with 380 nautical miles into Afghanistan in support of Operation Enduring Freedom."[384] The route flown on 12 and 13 April skirted the Turkish/Syrian border and led to a U.S.-run expeditionary airfield at Irbil, in northern Iraq, where the Marine expeditionary unit came under the tactical control of Joint Special Operations Task Force North. Although they welcomed the arrival of the expeditionary unit, General Osman and the Military Coordination and Liaison Command had no command relationship with the unit.[385]

The 26th MEU had been briefed for a mission to establish security checkpoints to prevent high-value targets, mostly senior Ba'athists, from trying to escape to Syria. One officer, First Lieutenant Sunny-James M. Risler, later recalled he had not known what to expect on arrival in Iraq; he was under the impression that he and his men needed to be ready to fight their way off the aircraft. But once they landed in Irbil, the Marines found that joint task force had the situation very well in hand; not only was the airfield secure, but there was a base camp complete with a good chow hall![386]

The only problem was that they were not going to stay long in Irbil. Soon after landing, Colonel Cleveland asked the Marines if they could take on a new mission—provide a stabilizing presence in Mosul. Both the battalion commander at the airhead, Lieutenant Colonel David K. Hough, and Colonel Frick were ready to adapt and respond, which pleased Colonel Cleveland, who was said to have been frustrated at times by other Services' lack of responsiveness. So, within a short period of time, the leading elements of 1st Battalion, 8th Marines, went on to Mosul without even having had a chance to find a good map of the place, let alone to plan. Over the next few days, the expeditionary unit flowed the rest of its assets to Mosul and established a presence around the local airport.

On the morning of 15 April, the 1st Battalion, 8th Marines, was involved in an incident that could have

A 26th Marine Expeditionary Unit explosive ordnance team, along with a security detail from 1st Battalion, 8th Marines, move in to search the remains of a demolished building near the airport in Mosul, Iraq.

Photo courtesy of Field History Branch

occurred almost anywhere in Iraq during this period. Tasked to conduct security operations for a regional coordination center in downtown Mosul, Company B drove to the government building, the former Ba'ath Party headquarters, along with civil affairs and Special Forces personnel. An angry crowd near the building became progressively angrier after a loudspeaker mounted on an ambulance began to broadcast a message in Arabic and a supposed "peacemaker" spoke to the crowd. It turned out that the crowd was angry because they believed that certain Iraqis who had benefited from Saddam's rule were to remain in power under the American occupation. Since none of the Americans at the governate that day spoke Arabic, they did not know what the problem was until much later. Some members of the crowd tried to punch at the Marines and to spit on them. Others tried to get close and, in the words of Sergeant Bryan L. Gilstrap, reach "into their pockets."[387]

The tension was palpable for the Marines, who were still getting adjusted to being in Iraq, to say nothing of a mission that was neither full-blown combat nor peacekeeping. True to their training, Marines checked the compound, manned the perimeter, and, according to Captain Colon, reacted to the first rounds that were fired at them by firing over the heads of the crowd. Sergeant Gilstrap, who was the scout-sniper team leader, remembered seeing an Iraqi raise an AK-47 rifle to a firing position, which made him a legitimate target under the rules of engagement. Gilstrap fired at, and hit, his target. This was followed by an intense firefight as the Marines and the crowd fired at each other. Sergeant Gilstrap said it was like the scene in the movie *Rules of Engagement* that showed Marines defending a U.S. Embassy in the Middle East against a mix of demonstrators and shooters (the movie raised questions about how much force the defenders were authorized, or entitled, to use). At some point a number of armed men tried to breach the perimeter and were killed or wounded by the Marines. As the unit's command chronology noted, the "crowd dispersed only after fixed-wing aircraft, coordinated through JSOTF-N, made low-flying passes over the area." There were no Marine casualties.[388]*

Two Iraqi teenage boys are unfazed by the presence of a Marine from the 26th Marine Expeditionary Unit, who is providing security at the Ibn Sina Teaching Hospital in Mosul.

JCCC 030420-F-5789F-005

Lieutenant Colonel Hough, the battalion commander, decided to keep his Marines at the site overnight, and the company spent the night improving its positions. The next day, the hostile sniping started back up. The Marines used a CH-46E helicopter as a counter sniper platform, with some success. But later in the day, Iraqi fighters began preparing to fire rocket-propelled grenades and crew-served weapons against the Marines from urban high ground. After consulting with his Army Special Forces counterpart, Colonel Frick decided to withdraw Company B. Given the force ratios and the general hostility in Mosul, it made sense to leave the exposed position at the government building. After dark, Company A and Special Forces soldiers ran a convoy of trucks into the site to return Company B to friendly lines. Although the evolution occurred without further incident, officers in the expeditionary unit operations section commented that they would have felt more comfortable running light armored vehicles or amphibious assault vehicles into the heart of town, which they had not brought with them into Iraq.[389]**

*Capt Arnaldo Colon reported seven confirmed kills. *The New York Times* reported that on 15 and 16 April there were 17 Iraqis killed and 39 wounded but does not specifically state that all of these casualties were from firefights at the government building. The command chronology does not state a precise number, only that "several Iraqis were killed or wounded." After the firefight there were allegations that the Marines had overreacted, that is, that their response had not been proportional to the threat. At Col Frick's behest, 26th MEU's staff judge advocate, Maj Ian D. Brasure, conducted an informal investigation to determine if there needed to be a more formal investigation of the allegations. After talking to participants in the firefight, he came to the conclusion that that would not be necessary. (Capt Arnaldo L. Colon, ed., "U.S. Marines in Northern Iraq: A Certain Force," n.d. [Filed with 26th MEU(SOC) ComdC, Jan-Jun03, GRC, Quantico, VA]; 26th MEU(SOC) ComdC, Jan-Jun03; Maj Ian D. Brasure telephone conversation with author, 17May04)

**According to Gen Osman, who was apparently recording information that had been passed to him: "Col Frick decided to abandon the RCC [Ba'ath Party headquarters] that had been occupied based on suggestions by several local leaders. Late yesterday, the Marines guarding the RCC spotted several men establishing a machine gun position and carrying an RPG in a building situated about 200 meters from the RCC. In light of the firefight from the previous day and what appeared to be an upcoming fight, it was decided a more secure location was in order." (Gen H. P. Osman sitrep to Gen J. P. Abizaid, 7Apr03,[Copy in Reynolds Working Papers, MCHC, Quantico, VA])

JCCC 030423-A-7170X-001

Soldiers from the 101st Airborne Division arrive at Mosul Airport on 23 April, relieving Marines of the 26th Marine Expeditionary Unit of security and stability operations in northern Iraq.

For the remainder of its time in Mosul, which lasted until it was relieved by the 101st Airborne Division (Air Assault) in late April, the 26th MEU conducted a variety of security missions in and around Mosul, none quite so dramatic as the first, but still important and useful. The expeditionary unit also put its aviation assets to use for the mutual benefit of the Marines and the Army Special Forces. This was especially true of the Marine KC-130 at the unit's disposal, which again proved their flexibility and their worth. On 23 April elements of the 101st began relieving the Marines at the airfield, and the 26th MEU began the process of disengaging and redeploying. By 2 May, its redeployment to Souda Bay was complete. Colonel Frick's view, in retrospect, was that the unit (like the coordination and liaison command) had successfully bought time for the Coalition—a stopgap force that had helped to restore some semblance of law and order in a difficult operating environment.[390]*

As the 101st took over from the Joint Special Operations Task Force, the Military Coordination and Liaison Command contributed its insights on the situation. General Osman spent a good part of 22 April with the incoming commander, Major General David H. Petraeus, explaining his work and passing on his views on recent developments in northern Iraq. In particular, Osman described Mosul as "different than any other city in Iraq" with its "feeling of having been defeated vs. liberated." He believed a strong United States presence was needed to enable key leadership to "feel comfortable enough to come forward." He concluded that as soon as the new leadership was able to establish itself, "we will want to remove the United States presence quickly."[391] General Osman conducted similar briefs with, among others, the commander of the 4th Infantry Division, Major General Raymond Odierno, and representatives of the Office of Reconstruction and Humanitarian Assistance. Some members of General Osman's staff thought the office was better prepared for a reprise of Operation Provide Comfort than for the situation that northern Iraq actually faced. Since its chief, retired Lieutenant General Jay M. Garner USA (Retired), was yet another veteran of that operation, this was understandable. Once he had conducted out-briefs with all of these officers, General Osman determined his mission was at an end and had arrangements made for his group to fly home.

What was the significance of General Osman's mission? What did the Military Coordination and Liaison Command contribute with some measure of success to a situation that was already being handled by the U.S. Army Special Forces? Arguably, the presence of a general officer freed the Special Forces from having to focus so heavily on day-to-day political relationships, thereby enabling them to focus more on the military side of things. At the same time, the command accomplished its own mission. General Osman did not win a war. The coordination and liaison command did not receive or return fire, although it came close on some days. But it did help to prevent a small war, and it did contribute to the success of Operation Iraqi Freedom as surely as if it had maneuvered a Marine air-ground task force over northern Iraq.[392]*

*Col John P. Holden, deputy chief of plans for Sixth Fleet, stated that the reason for redeploying the 26th MEU after such a short period of time was that Coalition Forces Land Component Command decided that with the 101st Airborne Division on site, there was no longer any need for the Marines. (Col John P. Holden intvw, 9Jun03 [MCHC, Quantico, VA])

*The Military Coordination and Liaison Command evokes the utility of the kind of engagement that was practiced so well by officers like Gen Zinni when he traveled around the CentCom area of operations to establish relationships with local leaders. The military historians Murray and Scales have argued that this is the next level the U.S. military needs to reach. Just as the Marine Corps and its sister Services learned how to operate lighter and faster between 1991 and 2003, there is now a need to learn how to be attuned to the cultural contexts in which they operate, refining the ability to find and manipulate "non-kinetic," social or political, pressure points, which can often eliminate the need for "kinetic," or traditional military, interventions.

Chapter 10

A Marriage of Convenience: Cooperation Between I MEF and the British Division

Like the war in the north, the war in southeast Iraq had some of the character of a private war. It was conducted by some 16,000 colorful fighters who had been fighting small wars for centuries, did things their own way, got along famously with Marines, and even spoke English.

The habitual relationships were there. British and American forces had been cooperating in the Persian Gulf since Desert Shield and Desert Storm. Their air forces jointly enforced the no-fly zones. After September 2001 there had been British liaison officers at CentCom and joint operations in Afghanistan. Some senior American and British decision makers already knew one another, which made planning that much easier, and there were a number of established joint procedures. The shared knowledge of the area, and of each other's procedures and personalities, made it easier for the British to succeed even though it was very late in the day when they started appearing in Kuwait in significant numbers to join I MEF.

Some of the unknowns about British participation were resolved only during the combat phase. The threshold issue was whether the British would fight alongside the Americans at all. Although the British prime minister, Tony Blair, proved to be a staunch ally before, during, and after the war, he faced a good deal of political opposition to his Iraq policy at home. This introduced an element of doubt, almost up to the last minute before G-Day, as to whether the British government would be able to order its troops into battle.

Even assuming that London would give the word to fight, there were still other significant unknowns. One was where the bulk of the British forces would attack from. Until December 2002, there were the plans to introduce most of the British from the north, that is, from Turkey, alongside the U.S. Army's 4th Infantry Division. According to testimony given to the British House of Commons: "In the very early planning, the Americans had decided to attack only from the south. . . . Militarily it made more sense to . . . attack on two axes, because there was going to be congestion. That was suggested to the Americans, who seized it with both hands, and that is why there was thought . . . of putting an axis through Turkey."[393] The truth is a little more nuanced. General David McKiernan, for one, commented in December 2002 that he had "always wanted a supporting attack out of the north that makes Saddam Hussein . . . look in two directions at large conventional forces."[394]

If a British division had attacked from the north, the British contribution to I MEF would probably have been nothing more than the Royal Marine unit earmarked for an amphibious operation against the peninsula in the southeastern corner of Iraq.* Always considered a piece of key terrain, Al Faw was an obvious choice for an amphibious assault of some sort. It faced Kuwait across a bay, making it a perfect launching pad for short-range missiles against Kuwait City or Coalition forces packed into the desert camps north of the Kuwaiti capital. It also straddled two important waterways, the channel to Iraq's only deep-water port, Umm Qasr, and the Shatt al Arab, which ran between the Persian Gulf and the "capital" of southern Iraq, Basrah.

By late December 2002, in the face of continued Turkish intransigence, the British had begun to explore "the southern option."[395] That was the month when there were preliminary talks between General James Conway and the senior British Army liaison officer at Coalition Forces Land Component Command (CFLCC), Brigadier General Albert E. Whitley, which helped to lay the groundwork for British cooperation with I MEF. Around the same time, British military officers were attending various CentCom conferences and exercises like Internal Look, but they were there only as observers, not as participants.[396]**

By early January 2003, I MEF and British planners were working hard to make up for lost time. Switch-

*"It was always planned that the Royal Marine Commandos would operate in the south." As the British Marine commander, Brigadier Jim Dutton, explained: "It was going to be . . . just one commando unit, in conjunction with the SEALs." (House of Commons, *Lessons of Iraq*, Vol. 1, p. 44)

**At senior levels, British and American officials held increasingly detailed discussions about Iraq from the summer of 2002 onward. It appears there was a continuum, a series of lesser decision points, rather than one or two decisive conferences or agreements. (House of Commons, *Lessons of Iraq*, v. 1, pp. 32, 34)

Photo courtesy of CFLCC

Coalition senior commanders, including LtGen David A. McKiernan and MajGen Robin V. Brims, commander of the 1 (UK) Armored Division, conduct final planning for the seizure of the southern Iraqi city of Basrah.

ing from the northern to the southern option meant the British would have to change their force mix. Instead of a predominantly armor, mechanized package, suitable for traveling and fighting over long distances, it now made sense to put together a lighter package, especially since there might be more fighting in the built-up areas of Umm Qasr and Basrah. So the British division, whose formal title was 1 (United Kingdom) Armored Division, evolved into a hybrid of forces, the more or less independent units from various commands and regions being roughly comparable to Marine regimental combat teams: 16 Air Assault Brigade, 3 Commando Brigade, and 7 Armored Brigade (the only one that was an organic part of the division). The Commando Brigade was the parent unit of the battalion-sized Royal Marine commando that had been aimed at Al Faw; it was now reinforced by another Royal Marine commando, but since it was still short of its full complement, the resulting gap was ultimately filled by the 15th Marine Expedition Unit.

The headquarters elements arrived in theater first, and, aware that CentCom was planning to be ready to fight sometime in March, did a remarkable job of orchestrating the surge of forces into Kuwait, running them through the CFLCC reception, staging, and integration process and establishing not only internal but also external command and control mechanisms. The man who made this happen was Major General Robin V. Brims, the division's commanding general, who was tall, thin, articulate, approachable, and full of energy. A career officer who had been commissioned into the light infantry and subsequently served in Germany, Northern Ireland, and the Balkans, he was the kind of soldier Americans did not meet every day. There was a trace of old-school eccentricity in his official biography, which described him as "single, a cricket fanatic, Newcastle United supporter, and outdoor enthusiast whose specialty area is bonfires."[397] It was a trait that made him all the more attractive to Marines, whose values sometimes seem to be from another era. In this he was not unlike General James Mattis, with whom he was to forge a good working relationship.[398]

Quick to visualize the desired outcome, General Brims put his stamp on the division, issuing his "GOC [General Officer Commanding] Directive 1" on 3 February 2003:

> We are TaCom CG I MEF. We create tactical effects to enable decisive delivery of his plans. We are integrating with HQ I MEF and its subordinate formations. Whilst we have been planning within I MEF for barely a month, much of I MEF has been planning, even conducting scenario related training, for many months. . . . We have identified five critical elements of 1 (UK) Armd Div's AO upon which we need to have an effect:
>
> a. Iraq's Armed Forces
> b. Irregular forces working against the Coalition
> c. Oilfield infrastructure
> d. UMM QASR
> e. BASRAH . . .

> We must also establish all our personal and electronic connectivities and processes with HQ I MEF. Our force is designed to be supported by I MEF deep assets. We must establish the techniques for delivering this in a timely, effective, and safe manner.
>
> We are maneuvrists. We must grasp opportunities to deliver our missions with minimum kinetic force. Iraq must still exist after the conflict as a sovereign state, stable and able to defend itself. . . . We shall probably be the first Coalition forces to implement Phase IV. We can set the pace.[399]

Despite their late arrival, the British received a warm welcome from I MEF. To review the bidding, I MEF could now assign some of its missions to the British and would not be stretched as thinly as it would otherwise have been. It could even ratchet the tempo of operations up still higher. With respect to the "Opening Gambit," I MEF decided not to change the carefully rehearsed plans to seize the oil fields in the first few days of the war. The planners considered the late date, and the extensive rehearsals that 1st Marine Division had already conducted, and decided it made more sense to leave well enough alone. The Marine division would still seize the objectives, but the British units, following close behind, would take over the security mission soon after the initial assault. Similarly, 15th MEU, operating under British control, would seize Umm Qasr but then turn the city over to the British and revert to U.S. Marine control.[400]

The Marines were so pleased with the presence and the potential of the British that by early March, they had decided to expand their area of operations by a further 250 kilometers to the north. According to the British division's war diary, written in somewhat formal English, "it was envisaged that after the execution of the initial Base Plan, which sees the division seizing the Al Faw Peninsula, subsequent operations would involve a push north by a brigade to the vicinity of Al Kut and Al Amarah. This would be conducted in parallel with 1st MarDiv and TF Tarawa to secure and screen the eastern Flank. The origins of this increased role apparently came from a mutually registered desire at all levels to maximize the full potential of the division's combat power."[401]

General Conway remembered later that one of the deciding factors in the distribution and timing of tasks was the amount of equipment available to the British as G-Day approached. The deployment phase was completed by 18 March, but the division was not ready to cross the line of departure. Quite simply, the U.S. Marines had more at their disposal, while the British would not be fully equipped until after the war had begun.[402] It was, as the British parliamentary report concluded, a "close run thing."[403]

By G-Day the division itself was combat ready; that is, ready for what in American terms was a running start, even though some of its units were not ready. It could carry out the missions that were to occur on the first days of the war, and would have to continue to prepare for the missions that were to occur after the initial battles. General Brims explained: "Sometimes you have to be positive . . . you are dealt a hand of cards . . . you would like to have 52 cards in the deck. . . . But, in this case, we had actually declared readiness with 46 cards."[404] His subordinate, Brigadier Graham J. Binns, the commander of 7 Armored Brigade, was not quite so sanguine, saying he "felt that we were carrying a lot of risk. The mood was one that we were not ready. . . that the soldiers . . . were not properly equipped."[405] But throughout the process there was a great deal of creativity and adaptation that made the British ready to cross into Iraq on 20 March.

The synergy that developed between the British division and I MEF between January and March 2003 certainly eased the joint preparations for combat. Over the years, the U.S. Marines had nourished a relationship with the Royal Marines, while the British Army had tended to forge its relations with other armies on the continent, and the Royal Air Force had cooperated with the U.S. Air Force in the Persian Gulf, all good preparation for fighting as part of a Coalition. It was something new for U.S. Marines and the British Army to work closely together, but it did not take either side long to discover the common ground. The two forces had similar cultures, traditionally oriented to "small wars," independent operations, and mission orders. The leaders' personalities meshed, not only between those of the same rank, as in the good relationship between Generals Mattis and Brims, but also between superior and subordinate. General Conway's command style sat very well with General Brims, who said that throughout his association with the Marines, he consistently felt empowered and never constrained.[406]

The process was strengthened by the exchange of liaison officers who actively interpreted their charters. The head of the U.S. liaison team to the British headquarters was Colonel Thomas C. Latsko, who embedded most of his Marines in the U.K. division's operations section, where they became active par-

JCCC 030604-M-865811-015

Before heading the liaison team with the British headquarters, Texas-native and Miami University of Ohio graduate, Col Thomas C. Latsko, had served as operations officer, 3d ANGLICO, battalion maintenance officer, 4th Tank Battalion, and as an advisor to the 1st Naval Construction Regiment.

ticipants in planning and operations. The Marines found that the British division was far more map-centric than any American command. Battle updates took place around the "bird table," a large table covered with maps and overlays for what looked to American eyes like World War II briefings. No one used PowerPoint software. The demeanor of the British officers around the table seemed informal, even casual; for example, there was little apparent consciousness of rank, and first names were used freely. But there could be little doubt about their professionalism. Units, and individuals, typically reported only what the next higher echelon needed to know. Briefers were held to time limits. Most British officers had a strong ability to visualize the situation, enter into creative and occasionally pointed debate about possible courses of action, and issue clear mission orders, which were far less detailed than American operations orders but, like most Coalition orders in this war, focused first on the desired effect, as opposed to simply directing a particular action.[407] Marines working with subordinate British commands made similar observations, saying they found British briefs to be straightforward, person-to-person, and not cluttered with detail—overall, a "nice change of pace."[408]

Colonel Latsko and his deputy, Lieutenant Colonel Edward J. Quinonez, made it their business to bring the British command-and-control system closer to the American one, and holding school at the British command center, introduced their new comrades in arms to various computerized American systems. By the end of the fighting in Iraq, the two systems were literally operating side by side, and British and American staff officers could work both at the same time to mutual benefit, although the gap between them, at least in terms of hardware, remained significant and was the subject of a British lessons-learned article.[409] One significant technological issue was SIPRnet connectivity. This internet channel was arguably the primary means of communication for U.S. forces in this war. Without access to SIPRnet, the British often felt at a loss, and liaison officers had to develop work-arounds.[410]

Alongside Colonel Latsko's officers in operations was a U.S. Marine air support element, which was able to provide the rapid response and coordination of close air support that is the function of the Marine direct air support control center. The air support element operated mostly out of the British combat operations and information center, built around the "bird table," but also sent its Marines down to the two British brigades that wanted their services. There they joined forces with Reserve Marines from the 1st and 3d Air-Naval Gunfire Liaison Companies (1st and 3d ANGLICOs), small numbers of whom had been deployed to the British division and immediately integrated into its training cycle. Recognizing their usefulness in joint and combined operations, General Jones had wisely brought the air and naval gunfire liaison companies back on line during his tenure as Commandant. This was proof of concept.[411] The 7 Armored Brigade had made a point of getting its air and naval gunfire liaison Marines to participate in training on British simulators in Germany, where they also made a point of exploring local drinking establishments together. As British officer Lieutenant Colonel Nicholas D. Ashmore commented, if shedding blood is the best way to build comradeship, drinking together is the second best.[412] The officers of 7 Armored viewed Marine air as the great "get out of jail" card, but thought that air and naval gunfire liaison companies alone would be enough and passed up an opportunity to have its own air support element, a decision it came to regret once its sister brigades discovered what a useful and flexible tool Marine air control was.[413]*

For the British the war began in much the same way as it had for I MEF headquarters. On 20 March, one of the incoming Iraqi theater ballistic missiles flew directly over the division headquarters, certainly capturing the attention of the staff. With British understatement, and humor, General Brims commented at the evening update brief that the "Iraqis were showing that they too had a vote" in how the war would run.[414] By the afternoon of 21 March, 3 Commando Brigade, the Royal Marine formation, had successfully deployed into the Al Faw Peninsula, despite "blackened skies from burning oil infrastructure" and a deadly crash of a U.S. Marine helicopter carrying Royal Marines.**

In the meantime, 15th MEU (SOC) took the port of Umm Qasr, encountering some stiff resistance, particularly from *Saddam Fedayeen*, but not enough to prevent it from making "excellent progress."[415] The next day, 3 Commando Brigade reported Umm Qasr "clear though 15 MEU continued to have sporadic contacts," while 7 Armored Brigade and 16 Air Assault Brigade each carried out a relief in place with the 5th and 7th Regimental Combat Teams, respectively, in the Rumaylah oil fields. There was some friction, a blue-on-blue "friendly fire" incident with no casualties between 15th MEU and 7 Armored Brigade on 21 March, and some confusion "due to the large volume of Coalition traffic" when the British division relieved the American division in the oil fields, not surprising considering the volume of friendly forces passing through the area.[416] But overall, the first days of the war went very well, and the British were pleased with the situation.

The 15th MEU (SOC), which was to chop back to U.S. control on 25 March after being relieved in Umm Qasr on 24 March, had meshed almost seamlessly with the British brigade. The Marine unit commander, Colonel Thomas D. Waldhauser, found it to be a great experience "by design and by default." He commented that even though the expeditionary unit's ground combat element had more organic combat power, there was a lot of congruence in the way both sides were organized. The Royal Marines were true to their doctrine, giving the U.S. Marines mission orders and then giving them the leeway to execute those orders, which were fourfold: to annihilate the enemy who fought; to accommodate the enemy who capitulated; not to destroy the infrastructure of the city; and not to get bogged down within city limits. Given the opportunity to make a contribution under unusual circumstances—this was not your normal cruise, the Marines of the 15th MEU (which, Waldhauser noted, had never carried out an exercise but only real-world operations) accomplished the mission with gusto.*

Overall, the British were on plan and the Iraqis were not; in particular, the British had been able to seize petroleum and shipping infrastructure before the Iraqis could do much harm to it. While it would prove impossible to get the oil flowing again quickly, it was soon possible to get ocean-going ships into Umm Qasr, especially to unload large quantities of supplies for the expected humanitarian crisis. The Royal Fleet Auxillery *Sir Galahad* (L3005), the first ship with a humanitarian load, began to unload on 28 March. Considering the overall situation, the Ministry of Defense postwar study concluded that:

> Four days into the campaign the Iraqi 51st Division had been removed from its defence of the oil fields. The 3 Commando Brigade held critical oil infrastructure at Al Faw and the port of Umm Qasr. The 16 Air Assault Brigade held

* The value of the Marine contribution is suggested by the House of Commons report on the war, which stated that the British forces needed more practice and training when it came to close air support. To the same effect, Capt Arnold M. Kiefer of 1st ANGLICO, found that the British did not have as much experience with combined arms as did the Marine Corps and tended to view the employment of supporting arms in a sequential way. They did not use and deconflict all of their supporting arms options. Capt Kiefer added the comment that their light infantry skills were otherwise world-class. (House of Commons, *Lessons of Iraq*, Vol. 1, p. 63; Capt Arnold M. Kiefer intvw, 17Mar03 (MCHC, Quantico, VA)

**"42 Commando's insertion started badly in appalling visibility, made worse by blowing sand and smoke from fires started the previous day. Tragically, the Marine Ch-46 "'Sea Knight" helicopter carrying the headquarters of the Brigade Reconnaissance Force crashed. . . . With the cloud base dropping still further, the insertion was aborted, forcing the Brigade HQ rapidly to identify other aviation assets and plan a new insertion for 42 Commando at dawn, using RAF Chinook and Puma helicopters. Although the landing took place six hours late, onto insecure landing sites, and in some case miles away from those originally intended, all objectives were secured." There was apparently some bad blood over the U.S. Marine Corps' decision to abort, which may explain why the British decided to use their own assets. (Cmdr 1 (UK) Armored Division's Diary, entry for 21Mar03 [Copy in Reynolds Working Papers, MCHC, Quantico, VA]; Ministry of Defence, *Operations in Iraq: Lessons for the Future* [London, UK: 11Dec03], p. 12)

*Col Thomas Waldhauser also noted the 15th MEU's air combat element was detached while the unit was in Iraq. He would have preferred to keep his own aviation combat element, but apart from that had no complaints about air support. This tracks with Task Force Tarawa's experience; its aviation combat element had also been stripped away when it landed in Kuwait. (Col Thomas D. Waldhauser intvw, 14Apr03 [MCHC, Quantico, VA])

DVIC DM-SD-04-01657

Fire and smoke from a damaged oil and gas separation plant is visible in the background as elements of the 1 (UK) Armored Division move into southern Iraq.

> the vital oilfield at Rumaylah and threatened Iraq's 6th Armoured Division to the north of Basrah to such an extent that it could not interfere with Coalition operations. The 7 Armoured Brigade held the bridges over the Shatt-al-Basrah waterway to the west of Basrah [a canal that ran from Basrah to Umm Qasr to the sea, paralleling the much larger Shatt-al-Arab to the east]. . . . [T]his was the most crucial ground to hold in order to achieve the overall plan of protecting the right flank of the U.S. advance to isolate Baghdad.[417]

Like their American allies to the west, the British were finding that the regular Iraqi Army was not as much of a threat as had been feared but that there might be more of an irregular threat than had been predicted. On the evening of 23 March, General Brims spoke with 7 Armored Brigade about "the changing nature of our understanding of the conflict. . . . There was not to be a solely conventional battle, rather resistance especially focused on urban areas, and troops either bypassed or deliberately inserted behind the forward line of our own troops."[418]

The issue confronting the British was what to do next. Ultimately they would have to occupy Basrah, Iraq's second city and that was not up for discussion. As General Brims commented: "to remove a regime you cannot leave [it] . . . in control of an urban area."[419] So it became a matter of timing and method. As for timing, CFLCC did not want the British to get ahead of the rest of the force; General McKiernan had directed them not to occupy Basrah proper until he gave the word. He told General Brims that he wanted the fight for Basrah to wait until the Coalition had isolated Baghdad, because he did not want to risk a dramatic urban fight in the south that could drive Iraqis into the arms of the regime and make the overall mission that much more difficult.[420] General Conway's guidance was similar but keyed to I MEF's scheme of maneuver; he directed Brims simply to make sure that whatever he chose to do in Basrah, he should remember the paramount goal of protecting I MEF's flank.[421]

What Generals McKiernan and Conway wanted was the reverse of what the Ba'athist strongman in Basrah wanted. He was Saddam Hussein's cousin, Ali Hassan Al-Majid, commonly known as "Chemical Ali" for his murderous suppression of the Kurds in 1988 with chemical weapons. He had been equally brutal in his dealings with the Shia in the south in the 1990s after the Gulf War, and was one of the prime movers in the draining of the swampland north of Basrah, the ancestral home of the "Marsh Arabs" and their distinctive culture, in order to eliminate it as a haven for potential or actual rebels. The result was an ecological and economic disaster. Since Basrah's approximately 1.25 million citizens were predominantly Shia, he was feared far more than respected. At the beginning of the war, he presided over a mix of unreliable regular army units and more reliable but not particularly skilled irregulars like the *Saddam Fedayeen*. To the extent that he had a strategy, it appears to have been one of trying to draw the British into a drawn-out fight within the city limits of Basrah, with the attendant collateral damage to cultural and religious sites, and of course civilian casualties, the kind of thing that hurts the Coalition when it appears on the front page of an Arab daily, not to mention *The Washington Post* or *Le Monde*. British intelligence reports suggested just how basic the Iraqi plan for Basrah was: "Whenever the British intercepted enemy communications, Saddam's henchmen were merely urging loyalists to fight, fight, fight, without specifying how."[422]

This was exactly what the British were determined not to do, whatever the timing of the attack. British officers commented repeatedly that they were always determined to avoid the kind of bitter, costly, house-to-house, street-to-street fighting that had recently occurred in Chechnya, where the Russian army had become bogged down in its fight with rebels, or like the Eastern Front in World War II. "We were determined," Brims commented after the war, "not to have

JCCC 030328-N-3783H-479

Royal Fleet Auxiliary Landing Ship Logistic, Sir Galahad *(L 3005), arrives at the Iraqi port of Umm Qasr to deliver the first shipment of humanitarian aid from Coalition forces.*

any sort of Grozny or Stalingrad scenes."[423]

Over the period between 25 March and 6 April, through a process of thinking and experimentation, almost like a Warfighting Laboratory evolution, the British came up with the alternative approach that fit the situation. The preliminary experiment took place in the much smaller city of Az Zubayr, population about 100,000, to the southwest of Basrah in 7 Armored Brigade's zone of action. The soldiers of 7 Armored, who wore the "Desert Rats" patch their predecessors had worn at the battle of El Alamein in World War II, found themselves taking heavy machine gun and rocket fire from irregulars every time they came near the city. The 7 Armored had the combat power to enter and reduce the city, but exercised restraint, starting out small with a raid into the city by 1 Black Watch battalion on 25 March. On 26 March, 7 Armored proceeded to isolate the city, which was not the same as besieging it but rather a matter of controlling ingress and egress or, just as important, demonstrating its ability to do so.[424*]

General Brims tells the Basrah story in a very English way as if he and his subordinate, Brigadier Binns, had "casually" solved the problem in the same way a Londoner might "casually" solve the wickedly convoluted *Times of London* crossword puzzle, when in fact it was their lifelong devotion to the art of war that was manifesting itself:

> I talked to the Brigade Commander . . . about four or five days into the thing . . . and he said, "I am going to work out how we are going to take Az Zubayr," and I said, "Good, I will go away and consider Basrah." And he said, "I have got the most powerful armoured brigade the British Army's ever put in the field, and I will back-brief you on my bit, of Az Zubayr, tomorrow morning." I arranged to see him first thing . . . and he asked me to come aside of him for a short time, and he said to me, "I have worked out, we cannot go into Az Zubayr . . . because that is what the regime want; we will inflict undue casualties, we will take undue casualties, we will hurt the civilians . . . that is what he is after. We have got to do it in a more cunning way." I said to him, "Well, that's funny, because I have worked out precisely the same thing for Basrah."[425]

On 26 March, General Brims convened a commander's conference at Brigadier Binns' headquarters, where they discussed the issue among

*As Gen James Mattis had demonstrated outside Al Kut, I MEF was very much aware of the legal implications of besieging a city, which both the Marine and the British divisions wanted to avoid.

themselves. Brims took the opportunity to announce the decision to make Basrah a divisional responsibility rather than delegating it to one of the brigades, as British commanders were more likely to do. The UK division would hope to foment popular unrest by conducting "deep operations" through the destruction of key targets while at the same time providing an opportunity for the people to receive humanitarian aid.[426]

What came next, in both Az Zubayr and Basrah, was a series of carefully orchestrated events, aimed as much at the enemy's mind as at his body. One of the prerequisites was the collection of the kind of citizen-by-citizen, neighborhood-by-neighborhood intelligence that can seem more like police work than part of a military operation. This was apparently something that the British government in southern Iraq had been pursuing for years. It was also a skill that the British Army had learned in Northern Ireland and the Balkans. By working with recruited agents, some of whom were reporting by cell phone, and by simply networking, the British were able to create a detailed picture of the life of the city, from the bottom up.[427]* Next came a round of carefully calculated raids, sometimes synchronized with raids by joint direct attack munitions, or smart bombs, or other precision munitions, which could lead to sniping or to "lodgements," something like the establishment of patrol bases that ranged in size up to the battalion level. For example, a British unit might stage a nighttime raid and then, when withdrawing, leave a sniper team behind to observe the neighborhood for a few days, or to snipe at Ba'ath and *Fedayeen* leaders.[428]** At the same time, information operations worked on the minds of the citizenry, by means of leaflets, broadcasts, or even targeted mobile telephone calls by General Brims himself. Then there might be a carefully planned air or artillery raid against a pinpoint target, whose effects the division could observe on the live feed from the "Phoenix," the British unmanned aerial vehicle. "The way we did it," General Brims reported to the House of Commons, "was to build up an intelligence picture, focused raids, ground raids, air raids, mind raids . . . [the people] wanted to be freed but they could not do it themselves, they needed our support, and therefore actually we had them helping us, and they were feeding us . . . accurate intelligence, worthy targets . . . and we were able to conduct these raids, and they had a very significant effect."[429]

DVIC DM-SD-04-01770

In Pinzguaer reconnaissance vehicles, British Army soldiers assigned to 2/1 Battery, 16th Air Assault Brigade, move out on patrol in southern Iraq.

General Brims made it clear he wanted some operations to occur simultaneously even though the focus of main effort would shift. Initially it would be on 7 Armored Brigade and Az Zubayr. After Az Zubayr was determined to have fallen on 4 April (there was of course no formal surrender ceremony) the lighter 3 Commando Brigade took over in Az Zubayr and 7 Armored moved on to augment the effort against Basrah, which remained a division-level operation. Resistance was becoming progressively lighter, while the population seemed to become more welcoming of the British.[430] As time went on and the British became more successful, they received more information from the citizens of Basrah. On 5 April they received a tip about the whereabouts of Chemical Ali, and the Coalition launched a smart bomb attack on the building where he was believed to be hiding. Although unsuccessful, it was believed to have been successful at the time and, at the least, a potent demonstration of Coalition power for all on the Iraqi side to see. Chemical Ali was not in fact captured until the summer.[431]

By now the U.S. Army and Marines were launching probes into Baghdad, and there was little chance that anything that happened in Basrah could disrupt operations in the north. The most recent raids having been deemed "very successful," General Brims ordered his division to execute "Operation Sinbad," the final push against Basrah, which was to come from a number of directions. It met with "a minimum of resistance from individuals with small arms" who were

*Murray and Scales make a comment to the effect that the British Intelligence Service had established a network in Basrah in the decade after the Gulf War, which if accurate suggests amazing foresight on the part of the British. (Williamson Murray , MajGen Robert H. Scales, Jr., *The Iraq War* [Cambridge, MA: Harvard University Press, 2003], p. 145)

**According to Murray and Scales, the regime loyalists were the ones with the cell phones; if you had a cell phone, you could be a target for a British sniper. (Murray and Scales, *Iraq War*, p. 149)

apparently either Ba'athist Party members or *Fedayeen*.[432] One objective after another, the sewage plant, the party headquarters, the governor's palace, fell without heavy fighting, something of which Brims was justifiably proud. Each success created the next opportunity, which Brims exploited; he commented later that his intent was to seek the right opportunity to execute each phase of the plan. By twilight on 6 April, most of Basrah was in Coalition hands, if not secure.[433]*

Murray and Scales recount a story about the next day that speaks volumes about the British Army's work in Iraq. On 7 April, the division committed paratroopers to the "old town" quarter of Basrah to finish off any lingering resistance. But the paratroopers found there was little for them to do and began to withdraw: "As the paras withdraw, Shia crowds began throwing rocks at British tanks and armored personnel carriers. One of the battle group commanders immediately sensed what was happening. He ordered his . . . crews, as well as the infantry, to get out of their vehicles, take off their helmets, stow most of their weapons, and walk out into the agitated crowd. Immediately the rock throwing ended and members of the crowd again smiled and clapped hands for the British troops."[434]

This remarkable story, which begins by reflecting Shia fears of abandonment by the Coalition, conveys something of the British sense for how to handle the transition to Phase IV, post combat operations. With his experience in Northern Ireland and the Balkans, the average British soldier may be one of the world's foremost experts on the three-block war. He could fight a conventional battle, defend himself against a guerrilla or paramilitary threat, and conduct humanitarian operations. He excelled at most of these tasks in southeastern Iraq. As early as 1 April, in Az Zubayr, British soldiers started to shed their helmets and patrol in berets. In Az Zubayr, Basrah, and, later, in

Photo courtesy of Field History Branch

As British troops, supported by tanks and armored personnel carriers, moved into Basrah, they encountered minimum resistance and the city fell without heavy fighting or loss of life.

Maysan province to the north, the same kind of bottom-up intelligence gathering that served the British well in combat was very useful in the transition period. Nevertheless, there was only so much this small force could do among a population of millions.

Like their American counterparts, the British did not have a definitive plan for Phase IV other than to be ready to handle a humanitarian crisis which did not develop anywhere in Iraq, and to hand off to the Office of Reconstruction and Humanitarian Assistance, which had focused much of its planning toward the same end. The British government as a whole and the British military in particular were in step with the rest of the Coalition in assuming that "post conflict . . . there might be humanitarian or environmental disasters of various sorts, refugee flows, shortage of food . . . and those kinds of issues. . . . ORHA was really designed, as far as we [British] could see, to prepare mainly for humanitarian issues."[435] Another similarity was their response to the looting that occurred in their zone. Like the Ameri-

*According to Murray and Scales, *The Iraq War*, "the initial plan of attack was to punch deep into the city and pull back out at night. . . . [But] the operation on this 'terribly long day' was going so successfully that Major General Brims decided to finish off the Iraqis with a final stroke." To similar effect, the British liaison officer to I MEF, Maj Simon Plummer, stated that what started out as a probe became a four-pronged, final assault when it became clear there was only minimal Iraqi resistance. While both sources are certainly consistent with the British practice of exploiting opportunities, as opposed to "fighting the plan," the firsthand sources suggest a slightly more deliberate, planned approach. As Gen Brims commented on 10 May, the division had developed a plan and waited for the right opportunity to execute it, as opposed to simply reacting to events. (Murray and Scales, *Iraq War*, p.151; Sudarsan Raghavan, "British Take Most of Basra," MiamiHerald.com, 6Apr03; MajGen Robin V. Brims intvw, 10May03 [MCHC, Quantico, VA]) cans, the British expected there would be some residual Iraqi police or army structure with which they could cooperate, and they had considered the possibility that there might be some looting, but they were taken aback by its scope and breadth.[436]

As in most other areas in Iraq, once the fighting stopped virtually all levels of Iraqi government in and around Basrah simply ceased to exist. The looters stepped into the resulting power vacuum and went to work with a vengeance in southeast Iraq, "they ransacked schools, hospitals and took away things . . . beds, chairs, and so on, or they just wrecked things."[437] The House of Commons Defence Committee judged that "the impact of this looting on the task of post-conflict reconstruction has been enormous," in terms of testing the goodwill of the people, making it difficult for them to return to work or school, get health care, or even accomplish the mundane tasks that local government performs for its citizens.[438] They were not prepared to cope with it; even if they had had the will to do so, they clearly lacked the numbers and, except in a few dramatic cases, did not intervene. The I MEF situation report for 7 April contained the laconic comment about Basrah: "looting ongoing, looters are only engaged if looting arms depots."[439] The committee concluded that the scale and shape of the force provided were best suited to achieving the Coalition's desired effects in the combat phase, but not to carrying those effects through into the post conflict phase.[440]*

A soldier from the Royal Logistics Service Battalion, 10th General Service Regiment, aids local Iraqis in the distribution of water. Elements of the battalion made daily water and food runs to Basrah and surrounding villages from the port of Umm Qasr.
JCCC 030415-D-1517P-002

In the spring of 2003, the situation in southeast Iraq did not seem as bad as the House of Commons and others would later say it had been. The forces on the ground were generally upbeat and believed they were making good progress. Examples from the division commander's diary include, on 9 April, the observation that "the mood in the city is still jubilant and . . . the buses continue to run"; on 12 April the report that 16 Air Assault Brigade was able to proceed north to Al Amarah in Maysan province for the relief in place with Task Force Tarawa without incident; on 13 April the fact that "joint patrolling with the local police force commenced . . . in Basrah"; and the conclusion on 21 April that "the local population are becoming increasingly committed to policing themselves and preserving their own resources." Shortly thereafter, on 22 April the United Nations declared the British zone permissive, a few days before President Bush made his declaration that major combat operations had ended. For the British this was the end of the combat phase; parts of the division, to include General Brims, made preparations to rotate home, with a sense of having accomplished their mission.

One of this British general's last official acts before leaving Iraq was to host a farewell reception for I MEF commanders and staff at his headquarters in Basrah International Airport on 10 May to celebrate not just success on the battlefield, but also the bond between the Marine and his division. Feelings that went beyond the usual routine expressions of official goodwill were on display. Senior Marines took the trouble to fly in from other parts of Iraq on KC-130s and went into the reception area in the airport's "VIP" quarters, which were opulent beyond belief or good taste, with gilded trimmings, high ceilings, marble floors, and thick carpets, but no working plumbing. Apart from the ultramodern Marine pattern desert digital utility uniforms, and the unusual British desert camouflage uniforms, it could have been a scene from the movie *Patton*, as white-coated stewards served gin and tonics to tired but happy officers who felt they had something to celebrate. The division band, very British and old-school, beat retreat as the sun went down, as if trying to close a chapter in history.

*This led into a somewhat philosophical discussion of whether the division's actions and effects desired by the division were really in sync. The point was that the division had won the war, but was anyone on the British side set up to win the peace?

Chapter 11

Postlude to Combat: Marines and Occupation of Iraq, 2003

Occupation was not a mission that Lieutenant General James T. Conway relished. He wanted his Marines to fight the war and then to "recock," to get ready for the next war. This was the pattern that came naturally to him and to many other Marines; the idea was to assault the beaches, seize the objective, and then move on to prepare for the next assault, leaving the occupation duties to others. When he discussed the issue with a journalist before the war, General Conway enumerated some of the issues that an occupier would face in Iraq, including some no-win choices, such as whether to intervene when Iraqis turned on one another after Saddam's fall. He concluded: "If I had a vote, I'would say let us get [I MEF] out of here." But he was quick to add that the Marines would probably have no choice but to involve themselves in the postwar occupation of Iraq in some form.[441]

There followed the decision for I MEF's future operations and plans officers to think about Phase IV and to draft an operations plan, even before the shooting war started. One of their starting points was liaison with Coalition Forces Land Component Command (CFLCC) planners and their counterparts at 1 (UK) Armored Division. By design or default, CentCom had left much of the responsibility for Phase IV planning to CFLCC, a dramatic change from its approach to Phases I-III. The I Marine Expeditionary Force turned to the British because they had some recent experience, as well as a useful staff study on the restoration of law and order after combat. The State Department's comprehensive "Future of Iraq" study, which laid out many of the challenges the United States would face in Phase IV, was considered taboo for military planners, because it was not compatible with Pentagon policy-makers' vision of postwar Iraq.

Retired U.S. Army LtGen Jay M. Garner, head of the Office of Reconstruction and Humanitarian Assistance, greets arriving delegates to the Iraqi Interim Authority Conference in Baghdad. The authority was to govern Iraq until formal elections could be held.

JCCC 030428-F-5918G-016

DVIC DF-SD-05-04667

U.S. Navy UT1 Chyne Greek, left, from Naval Mobile Construction Battalion 7, U.S. Army LtCol Matthew Gapinski, of the 358th Civil Affairs Battalion, and Maj Robert V. Carr, Civil Affairs team leader with the 15th Marine Expeditionary Unit, assess damage to the water treatment facility located at An Nasiriyah, Iraq.

Coalition Forces Land Component Command also had a loosely defined relationship with a group known as Combined Joint Task Force IV (CJTF-IV), so poorly funded that its members had had to attend trade shows to obtain office supplies, and with what was, in effect, CJTF-IV's successor organization, the Office of Reconstruction and Humanitarian Assistance. Both organizations were more or less under the tactical control of Lieutenant General David D. McKiernan, but it was no secret that office's chief, retired Lieutenant General Jay M. Garner, had direct lines of his own to the Pentagon.[442*]

Coalition Forces Land Component Command came to assign nine governates (roughly equivalent to provinces) in southern Iraq to I MEF, covering a territory three times the size of Virginia with a population of 9 to 10 million people, and eventually issued a formal plan known as "Eclipse II" to guide its subordinates in Phase IV. ("Eclipse I" had been the plan for the occupation of Germany after World War II.) Land Component Command focused on security and emergency repairs to the infrastructure. The assumption was that parts of the Iraqi government would still be in working order when the shooting stopped, able to maintain the infrastructure and that the Office of Reconstruction and Humanitarian Assistance would be able to step in to take on many tasks, especially those dealing with humanitarian assistance. As General McKiernan's special assistant, Terry Moran, commented, CFLCC planned to "leverage the Iraqi Army and . . . the [Iraqi] bureaucracy."[443] The Pentagon's original assumption was that even reconstruction and humanitarian assistance office would not have to conduct operations for more than a few months in Iraq, and that after a relatively brief occupation the Coalition could turn the country over to an Iraqi government.

Within the framework of Eclipse II, I MEF had a considerable amount of leeway, and its planners developed their own approach to the matter. First they did an "intelligence preparation of the battlefield," looking at the tribes in the various regions, their infrastructure, and the various threats Marines were likely to face. They came up with two assumptions: that the first 6 to 12 weeks would be critical, this was when precedents would be set, and that it was important not to try to do too much. The Marines had to keep from involving themselves too deeply in local affairs and to let the Iraqis solve as many of their own problems as they could. One of the future operations officers, Lieutenant Colonel Brian K. McCrary, remembered the many and varied unknowns that he and his counterparts discussed: how to vet

*Gen Tommy Franks wrote that he left it to the Pentagon to plan and run post combat operations and noted that Office of Reconstruction and Humanitarian Assistance lacked the resources it needed to do its job. Franks' decision was later criticized. (Franks, *American Soldier*, pp. 441, 524, 526)

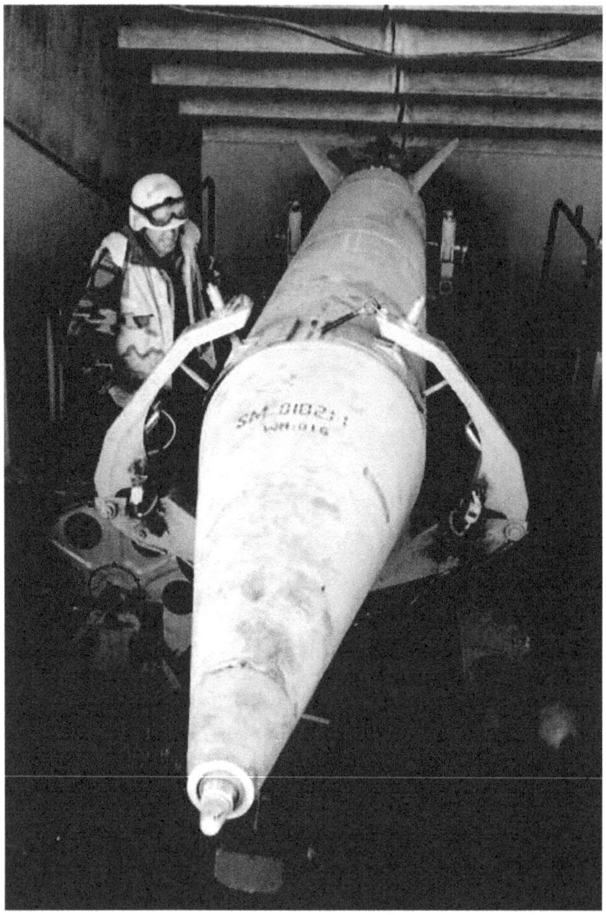

Photo courtesy of CFLCC

BGen Richard F. Natonski, commanding general of Task Force Tarawa, examines an Al Samut missile that was found at the Amara Soccer Stadium.

and pay local police; how to provide routine municipal services (power and water); how to find, secure, and exploit suspected weapons of mass destruction sites; how to secure the many weapons caches from looting; how to secure and process mass grave sites of victims of the regime; even how to run prisons.[444] This list did not include searching for Saddam Hussein and his sons, along with other prominent members of the regime; such high-priority tasks were taken for granted.

The interim result of the planners' work was a scheme for dividing the various responsibilities among the 1st Marine Division, Task Force Tarawa, and the British division, which declared it did not need much U.S. assistance with Phase IV. During a briefing to General Conway in early March, the planners provided him with options for "transitioning the force" and "enabling" the subordinate commands, especially by way of groups that would be known as "governate support teams," small teams of experts that could deploy to a locality. As plans officer, Lieutenant Colonel George W. Smith, remembered thinking, it was a good beginning; I MEF had defined the problem. But, he felt, the planners had only been able to point in the general direction of a solution but not to lay one out in detail.[445]

General Conway announced on 15 April that it would soon take up its postwar stance in southern Iraq. The announcement contained the Marines' trademark tinge of remorse about even having to conduct Phase IV operations. Lieutenant Colonel Smith, who doubled as the I MEF spokesman, explained to the press: "It is a tremendous responsibility and it's very complex. We focus the majority of our efforts on war-fighting. That is what we do. And so post-hostilities introduce a whole new spectrum of challenges. . . . We see that [fighting remnants of the *Fedayeen*] . . . as the number one threat. . . . We are going to aggressively hunt these guys down and . . . destroy them."[446]

The next day, General Franks convened a meeting of his senior commanders at Saddam Hussein's Abu Ghraib Palace in Baghdad, "an extravagant amalgam of marble, tile, gold fittings and massive chandeliers, all surrounded by an azure moat," to seal the Coalition's victory over the dictator and to ratify the plans for securing the country.[447] The Marine expeditionary force apparently held its commander's conference on Phase IV on the same day, and division followed suit shortly thereafter.[448]

With input from the Army's 358th Civil Affairs Brigade, which remained a welcome adjunct to the Marines during the occupation phase, I MEF published its order for Phase IV four days later. The planners' assumptions and the Marines' preference for keeping the occupation short and looking to the future after leaving Iraq, are clear from the text:

> On order, I MEF transitions to security/stability operations–establishes military authority, defeats remaining Iraqi combatants, maintains Iraqi territorial integrity, secures WMD [weapons of mass destruction] in sectors, and supports humanitarian assistance [and] the restoration of Iraqi civilian administration/infrastructure IOT [in order to] . . . enable a rapid transition to follow-on Coalition forces. . . .
>
> A guiding constant, to "enable," will be the basis for all that we do. We must enable IOs/NGOs [international organizations/nongovernmental organizations] and follow-on Coalition partners to support the Iraqi people. . . . We must enable the Iraqi people to support and govern

JCCC 030511-M-9792P-018

LtCol Daniel O'Donohue, commanding 2d Battalion, 5th Marines, speaks with local officials of An Samawah about what can be provided to rebuild the town.

> themselves. . . . Our overarching focus will be establishment of a secure environment as we disarm remaining Iraqi forces, . . . [re-create] local police forces . . . and develop . . . [a small new] Iraqi military. . . .
>
> We must clearly communicate to the Iraqi people the temporary nature of our mission and our desire to quickly transition to Iraqi self-determination. . . . While accomplishing . . . Phase IV, I MEF must also look "deep" toward the requirement of reconstitution, regeneration and redeployment [of the force].[449]*

As planned, there had been a blurred transition to Phase IV. Task Force Tarawa and the British and American divisions were conducting Phase IV operations before the force published its order, in Nasiriyah, Basrah, Baghdad, and many localities in between. This was partly by design, and partly the result of chance. The rolling transition from Phase III to Phase IV was expected, but the speed and suddenness of the regime's collapse had been breathtaking. The shift was perhaps clearest for the wing. From one day to the next, without a great deal of warning, the kill boxes (the targeting control measure in the air war) were simply "closed." This apparently happened on 11 April. There would still be calls for air support of various sorts, but it would no longer be a routine occurrence, and the wing could start thinking about drawing down in theater. On the ground, the violence also tapered off throughout the country, especially in the Marine area of operations in southern Iraq, although the change for an infantryman or the driver of a light armored vehicle was a little less dramatic. He still went out on patrol, with his weapons loaded, and had to be prepared for a fight at any time.[450]

While the British division ran its own occupation, with some success, by all accounts, General James Mattis' division set the pace for the occupation in the Marine governates. He already had a distinct vision

*I MEF Fires helped to prepare the order; once the need for planning deep fires had passed, both I MEF and CFLCC used their skills as planners of "effects-based operations" for planning and assessing Phase IV operations.

I MEF and the Law of War: How Marines Treated Iraqis

Before I MEF crossed the line of departure into Iraq on 20 March, every Marine received clear guidance: the enemy was the Iraqi regime, not the Iraqi people. Starting with General Conway, commanders went to considerable lengths to disseminate, down to the frontline Marine, the rules of engagement and the CentCom mandate to limit collateral damage, along with their own views on avoiding "triumphalism" or disrespect for the people and customs of Iraq. Marines were to be liberators, not conquerors, and they were to obey the Law of War. This was not an afterthought, an addendum to other kinds of training, but a theme that ran throughout the expeditionary force's preparations for war. It was especially true for 1st Marine Division, the major subordinate command most likely to encounter the enemy face-to-face.*

The force's legal office spent the months before the war working up the rules of engagement and then preparing and disseminating presentations for the major subordinate commands. Following I MEF's lead, General Mattis issued written, detailed guidance on the Law of War on at least two occasions, in addition to continuously repeating and explaining the division's motto, "No better friend, no worse enemy." In one prewar memorandum, he predicted that Marine "discipline will be severely tested by an unscrupulously led enemy who is likely to commit Law of War violations," and went on to outline 11 commonsense Law of War "principles."** To the same effect, in late 2002 as he sensed the approach of war, he urged his commanders to prepare their Marines for the probability of asymmetric attacks and, at the same time, reminded them that "both decisive force and chivalry will be critical to freeing Iraq."***

To drive the message home, General Mattis had his staff judge advocate, Lieutenant Colonel John R. Ewers, and his deputy, Major Joseph A. Lore, deliver classes on the Law of War and Rules of Engagement to division units both before and during the deployment; the general wanted the message to come from an independent expert on the subject, not the unit commander. Mattis went so far as to have Ewers create a team to travel around the battlefield and investigate allegations of wrongdoing before the smoke had cleared. The investigation involved a report that a media vehicle had been hit by fire from the 1st Tank Battalion. Riding in a soft-skinned humvee, Ewers was doing just that when he was wounded on 23 March 2003 near the town of Az Zubayr in southeastern Iraq. ****

It is fair to ask whether the Marines followed this guidance before and after combat. At this stage, the evidence, recorded in situation reports and through contact with Marine lawyers, is largely anecdotal and may never progress much beyond that. It appears that by and large most Marines did as they were told, sometimes even going the extra mile for Iraqis, but some Marines occasionally departed from the spirit or the letter of the law.

Perhaps the best-known example of "triumphalism," which was nipped in the bud, was the incident at Firdos Square in which Corporal Edward Chin placed an American flag over Saddam's statue, but quickly replaced it with an Iraqi flag and left it to the citizens of Baghdad to complete the work of destroying that symbol of the regime. Other, less well-known examples have to do with the care many Marines took not to kill civilians, even when there were legitimate targets nearby. In one case on 2 April, which illustrates the dilemmas that conscientious Marines faced in Iraq, Major Peter S. Blake, an AV-8B Harrier pilot, waited for a gap in the civilian traffic, which had, maddeningly, continued to flow in and around the battlefield all over the country, before launching his attack against an Iraqi multiple rocket launcher with a "guided" bomb. Within the next few seconds, a civilian truck came into view, and Blake decided to "slew" (or misguide) the bomb into the Tigris River to save innocent lives. He waited again for a break in the traffic before launching a second, and final, attack. Even though at the last moment another civilian truck appeared, he felt this time he had no choice, since 5th Marines was almost literally around the corner.***** There are numerous other such examples, like the one in early April when Marines held their fire until they knew whether an approaching school bus was filled with enemy fighters or innocent civilians, which turned out to be the case.****** In another, more personal example, on 29 March members of the 1st Force Service Support Group took the trouble to bury a two-year-old Iraqi boy who had stepped on a landmine and been evacuated to Charlie Surgical Support Company, Health Services Battalion, then in the vicinity of Jalibah, where doctors tried in vain to save his life. The boy had come to the field hospital without identification or relatives, and the Marines and sailors who were present decided to give him a proper Muslim burial. They researched the subject carefully, and then found three Muslims to perform the ceremony in the prescribed tradition, which included wrapping the body in a shroud and placing it in a grave with the child's face toward Mecca. The chief of staff of the force service support group, Colonel Darrell L. Moore, took time out of his busy day to assist in the ceremony. This small act was one of decency, pure and simple; no journalists or angry townspeople were present to demand that the Marines "do the right thing."*******

On the other side of the ledger, a few sources record cases where Marines overstepped the bounds of fire discipline or military law. In December 2003, the *Marine Corps Times* reported on the case of eight Marines charged in the death of a 52-year-old Iraqi detainee who was found dead in his cell in An Nasiriyah in the spring of 2003. The charges included "willfully failing to properly safeguard the health" of those in custody, as well as assault and making false official statements.******** According to the deputy staff judge advocate of 1st Marine Division, Major Lore, this was one of a handful of similar incidents that was investigated and resulted either in nonjudicial punishment or court-martial. In their book *The March Up*, Major General Ray L. Smith and Bing West tell a story about the time when Brigadier General John F. Kelly, the assistant commander of 1st Marine Division, admonished an unnamed battalion commander on 26 March for letting his troops needlessly shoot at approaching civilians. The Marines had a legitimate concern about suicide bombers in civilian vehicles, but that did not mean every heedless civilian who came within half a mile of a Marine position had to be stopped with deadly force. Although West and Smith go on to discuss their impression that most Marines agonized over the decision to fire or not to fire, two journalists who spent time with different units during the combat phase of the war, Evan Wright and Peter Maas, convey the impression that some young Marines were all too ready to fire at civilians, and that their officers and noncommissioned officers did not always do a good job of restraining them. This was especially true after Marines learned about the fighting at An Nasiriyah, where the enemy had worn civilian clothes and Marine casualties were heavy. At that point, the enemy for many, if not most, Marines became any Iraqi with a gun.*********

It is important to note that at least five such cases resulted in some form of disciplinary proceedings against the perpetrators. According to Major Lore, these proceedings ran the gamut from nonjudicial punishment to general court-martial. With the passage of time, there will be more clarity on these events. As of this writing, even basic statistics are difficult to come by. The I MEF staff judge advocate's command chronology for the first half of 2003 states that during Phase III of Operation Iraqi Freedom its primary focus was reporting and investigating violations of the laws of armed conflict. The same document shows that the I MEF staff judge advocate was involved in some way in one general court-martial, eight special courts-martial, and one summary court-martial, but without delving into the individual cases it is difficult to interpret the significance of these statistics.

The bottom line? The I MEF commander and his subordinates did the right thing. They told virtually every Marine what they expected, and they did so in some detail. During the war, the Marine leadership took measures to enforce the standards it had set. Most Marines appear to have met those standards, while a handful fell short and were disciplined when a case could be made against them.

---

*For basic data on I MEF-level staff judge advocates, see I MEF SJA ComdC, Jan-Jun03 (GRC, Quantico, VA). See also Gen Conway's talks on the rules of engagement in Chapter 3 and Col William D. Durrett intvws, 11Feb03, 9Jun03 (MCHC, Quantico, VA).

**MajGen James N. Mattis, "Commanding General's Guidance on Law of War," u.d. (2002?) (Copy in Reynolds Working Papers, MCHC, Quantico, VA). It is interesting to note that one of the six books on Gen Mattis' reading list for his subordinate commanders was *Son Thang: An American War Crime*, by Gary Solis, which was about how the 1st Marine Division handled a war crime in Vietnam.

***MajGen James N. Mattis, "Memorandum for All Commanders," 20Dec02 (Copy in Reynolds Working Papers, MCHC, Quantico, VA). The division staff judge advocate, LtCol John R. Ewers, remembers having "a number of conversations with both Gen Mattis and Gen Kelly [the assistant division commander] . . . about law of war . . . and the challenges posed in . . . asymmetric warfare with a foe who was expected to . . . break the rules." LtCol John R. Ewers, e-mails to author, 6, 7Apr04 (Copies in Reynolds Working Papers, MCHC, Quantico, VA).

****Ibid.; Maj Joseph Lore, e-mails to author, 5Apr04 (Copies in Reynolds Working Papers, MCHC, Quantico, VA).

*****Maj Peter S. Blake intvw, 15Apr03 (MCHC, Quantico, VA).

******Col Randall W. Holm intvw, 31May03 (MCHC, Quantico, VA).

*******Maj Melissa D. Kuo, "Field History Journal," entry for 29Mar03 (MCHC, Quantico, VA).

********Gidget Fuentes, "Hearing Scrutinizes Reservists' Handling of Prisoners in Iraq," *Marine Corps Times*, Dec03, p. 10.

*********Evan Wright, "The Killer Elite," *Rolling Stone*, 13Jun03, 24Jul03; Peter Maas, "Good Kills," *The New York Times Magazine*, 20Apr03. Wright's articles pull few punches and tell the story of a slice of the war from the corporals and sergeants perspective. They were subsequently expanded into the book *Generation Kill* (New York, NY: Putnam, 2004), which, while it is well written, has had mixed reviews from Marines.

for the way he wanted to implement the force order. Like I MEF, 1st Marine Division would issue general guidelines. Each of its governates would be run by a reinforced infantry battalion. The reinforcements might consist of a governate support team, a psychological operations team, a human intelligence exploitation team, civil affairs elements, and sometimes engineer or naval construction elements. The battalion commander would have a great deal of autonomy, but given the small size of his force, he would have no choice but to rely heavily on the international and nongovernmental organizations as well as whatever Iraqis he could mobilize. Interestingly, he would work not for his regiment but for division itself; General Mattis' plan was for the regimental staffs to focus on the retrograde and on preparations for the next contingency, the "recocking" that figured in so many plans and talks. This would enable him to satisfy the twin mandates in the I MEF order, not to mention the spirit of the order.

It was a dramatic move. General Mattis began the occupation by sending about 15,000 of his 23,000 Marines home, along with all of his tanks and assault amphibious vehicles, his armored personnel carriers, but not his light armored vehicles, which remained very useful. This was one way to reinforce his message about how to occupy a country, both to his Marines and to the Iraqis. He still intended for the Marines to patrol on foot and for the two groups to become intimately familiar with each other, even to trust each other. But he did not want "a heavy boot print" or the "sense of oppression" that could come if "everywhere you looked you saw a Marine. If we needed more people . . . I wanted to enlist the Iraqis [for] . . . our common cause."[451]

The 1st Marine Division had three basic policies. The first was "Do no harm," expressed in the kind of Mattis aphorism that all Marines could understand: "If someone needs shooting, shoot him. If someone does not need shooting, protect him." The second was to win heart and minds through good works. The third and final was to be ready at all times to win the 10-second gunfight.[452] The general's statements of policy branched into specific guidance. Because he wanted the Marines and the Iraqis to trust each other, he literally wanted them to look each other in

Cpl Michael C. Brown of the 4th Light Armored Reconnaissance Battalion patrols the streets of Al Kut with a member of the local police force. The purpose of the joint foot patrols was to let the residents see the Marines and police working together in order to put a stop to the looting after the previous government was abolished.

Photo courtesy of CFLCC

the eye; in fact, one of his directives was for Marines to remove their sunglasses when speaking to Iraqis. Another piece of guidance followed from his commonsense observation that the Iraqis would cooperate more readily if the Marines helped their children, and he sent units into the local schools to clean them up and get them running again.

Again like the force, division began its work in each province with a survey, with a view to developing a campaign plan tailored to the needs of each province. Generally speaking, the routine was similar to the one that division had followed in Baghdad before I MEF had published its order for Phase IV. Reestablishing security was paramount. Marines themselves did some of that work, but wanted to vet and train Iraqi police to take over from them as quickly as possible. The Iraqis joined Marine-led patrols with a view to gradually switching roles.

The division's various locations included Al Muthanna (2d Battalion, 5th Marines); Karbala (3d Battalion, 7th Marines); Al Qadisiyah (3d Battalion, 5th Marines); An Najaf (1st Battalion, 7th Marines); and Babil (1st Battalion, 4th Marines). The provinces of Wasit and Dhi Qar were added when 2d Battalion, 25th Marines; 3d Battalion, 23d Marines; and the 4th Light Armored Reconnaissance Battalion came over to division from Task Force Tarawa, which was rotating home at the end of May.

The 3d Battalion, 5th Marines, was blessed with an unusual Iraqi partner who became the local police chief, Brigadier General Fuad Hani Faris. A wounded veteran of the Iran-Iraq War, he was said to be critical of Saddam and appears to have been one of the many Iraqi soldiers who simply wrote their own travel orders when the war began. Faris moved himself from Hillah, where he had been assigned, to Ad Diwaniyah, where his wife's family lived. When the postwar looting began, he organized Iraqi soldiers into guard forces and protected some of the factories, ammunition supply points, and government buildings in the city until the U.S. Army appeared. He happily agreed to help rebuild the area and transferred his loyalties to the Marines when they replaced the Army in Ad Diwaniyah.[453]

The experiences of the Weapons Company, 3d Battalion, 5th Marines, in Ad Diwaniyah in early May illustrate what it was like for junior Marines charged with policing a city. The company was quartered in town in a villa that had belonged to a recently departed strongman. Set in a large, pleasant, walled compound next to a stream and including a vegetable garden and a more formal garden, the airy, two-story villa had been looted in early April, stripped even of doorframes and floor tiles. The Marines had to begin by cleaning it up and restoring it. The idea was to turn it into a police station, complete with armory and holding cells. Next the company put out the word that it would be paying local policemen, which quickly attracted them to the small base and allowed the Marines to begin the process of vetting, organizing, training, and equipping them. In the words of one observer:

> This afternoon the Marines were going to pay the Iraqi [policemen]–and issue weapons. . . . The Marines frisked the Iraqis as they came in [to the compound], confiscated (temporarily) their AK-47s (the small version with the folding stock . . .) but let them keep their pistols if they were unloaded. Marines in HMMWVs overwatched the process and armed Marines stood near the Iraqis as they formed themselves up. There were traffic police–dressed in white, with [peaked] hats like those worn by police all over the world–and security police–dressed in green outfits with the [black] Iraqi berets–all worn in different styles from the pillbox (which looks ridiculous) to the usual Iraqi mushroom shape (which is not that snappy either). No one looked hungry–unlike some of the people on the streets. A couple of the . . . [policemen] bordered on obesity. There were a few officers among them–[men with more braid who looked like] they were used to being in charge and leading (manipulating?) others. I watched one of them work his wiles on one of the translators used by the Marines. By and large, it seemed like a good-natured crowd. A few of them . . . [were] a little nervous as I took photographs–but most smiled and waved at me.[454]

In the meantime Marines continued to patrol the streets, apprehend looters, and hold them for a day or two in a makeshift outdoor prison, performing various kinds of less than pleasant tasks around the compound, such as filling sandbags or burning waste in the latrines. Whether the looters learned their lesson was doubtful. One of the looters claimed he was innocent because he had been hired to loot, saying the man the Marines should arrest was his employer. One thing the Marines learned was that no one wanted to be a policeman in Ad Diwaniyah without a weapon, no one seemed to like policemen, who bore the double stigma of having worked for the former regime and now of collaborating with the occupiers, and so they quickly decided to allow the Iraqis

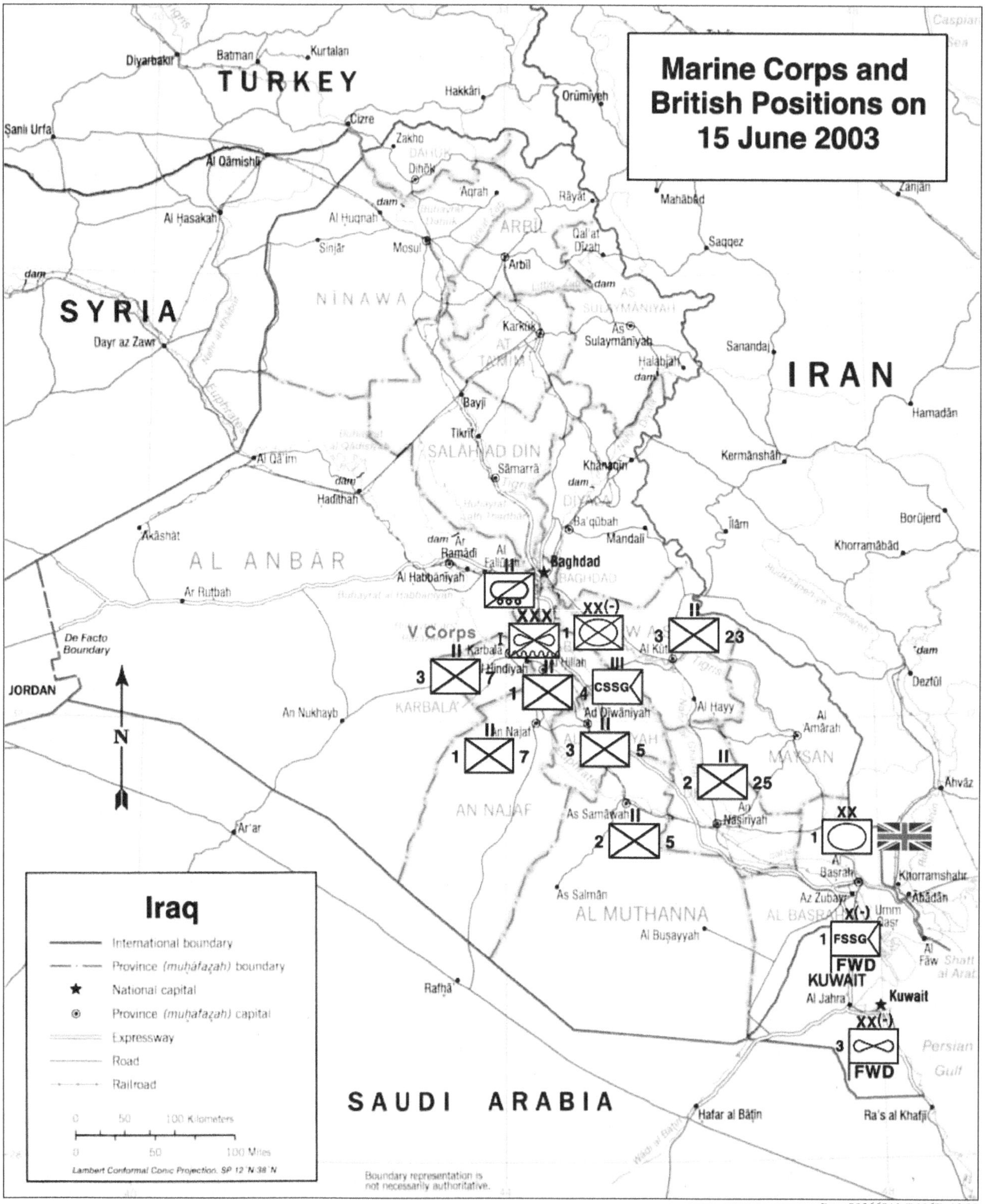

Base 802668AI (R0067) 12-99

to carry sidearms and AK-47 rifles.

While some Marines addressed security, others worked on the infrastructure. At the battalion command post, located on the grounds of a modern but unfinished medical school campus, there was a thriving Civil Affairs section and a chart showing the lines of operation, which included such entries as "Water/Sewage, Electrical, Medical, Education and Law, Fire, Public Transportation, Food and Distribution." An officer's name was written in next to each entry. For example, Second Lieutenant Glen J. Bayliff, whose main qualification for the job was that he was the logistics officer, was responsible for transportation, which meant conducting surveys, remaining abreast of developments, and helping to coordinate indigenous efforts with those of international or nongovernmental organizations. Sometimes Marine Reservists, described as the Corps' "hidden asset" by one grateful battalion commander, came to the rescue with their civilian skill sets as policemen, lawyers, or city managers. Many Marines found that neither the Office of Reconstruction and Humanitarian Assistance nor the organization that replaced it, the Coalition Provisional Authority, had enough resources to weigh in at the day-to-day, working level. For the most part, they were left to their own devices.[455]

Some national events took place that did have repercussions on the local level. Before the war, President Bush is said to have approved a plan that would have put "several hundred thousand Iraqi soldiers on the U.S. payroll and kept them available" for various tasks from providing security to repairing roads. But in a surprise move on 23 May, Ambassador L. Paul Bremer, head of the Coalition Provisional Authority, issued a decree disbanding the army and canceling pensions.[456] Some three weeks later, on 15 June, CFLCC turned over responsibility for the occupation of Iraq to V Corps, while keeping its responsibilities for supporting the force. This was the last day of the amnesty period under the Coalition decree, which limited the number and types of weapons Iraqis could possess. The predictable result of both policies was an upswing in violence against Coalition forces, especially in the area west of Baghdad around Ar Ramadi and Fallujah, two names that would become all too familiar to Marines in 2004.

This was not part of I MEF's area of operations, but it was contiguous to it. Army convoys traveling from Kuwait to Ar Ramadi and Fallujah had to pass through northern Babil Province, which did fall under I MEF. The Marines' future operations staff "began studying and planning to defeat this threat." The result was a plan to set up an armored task force to patrol the area.[457]

Built around the Armored Reconnaissance Battalion, under the spirited command of Lieutenant Colonel Andrew Pappas, the task force took the name "Scorpion" and was clearly undeterred either by the challenge or the marginal living conditions it faced. A visiting journalist, Pamela Hess of United Press International, captured the spirit of Camp Scorpion:

> Mad Max would turn up his nose. . . . There is nothing but garbage and dirt and sand as far as the eye can see. Marines live and sleep in the open air of a gravel parking lot, except for the few one-story concrete buildings that are air-conditioned on the rare occasions the generators can be coaxed to work. They have no chow hall . . . and until . . . recently . . . [sanitary] facilities were a plywood bench with four holes. . . . Powerful winds sweep the grounds, kicking up massive dust clouds that coat everything in dull brown powder several times a day.

The Marine

In April and May 2003, there was almost universal praise for the young enlisted Marines who served in Kuwait and Iraq. The historians who deployed from Washington, D.C., to the field to conduct interviews during and after the war heard story after story praising the spirit, ingenuity, and "stick-to-itiveness" of the individual Marine, and they often experienced it themselves. It was the small acts that stood out, whether it was the Marines who sensed, rightly, that spare parts would be in short supply up the road in Iraq and stopped to strip them from wrecked vehicles as they moved north, or the amphibious assault vehicle mechanics who repaired vehicles while they were moving under tow, or the Marines performing the lonely work of unwinding the hose reel in the desert, or the determination "to get there" of the driver of a seven-ton truck making his way through the sand at night, or caring for children under fire. Whatever his assigned task, each of these Marines was also prepared to fight as a rifleman, revalidating one of the basic tenets of the Marine credo. What struck many officers after the war was how well combat infantrymen adapted to the demands of occupation duty, switching from a "weapons free" to a "weapons tight" frame of mind. They marveled at the restraint that these young men and women were now showing. The praise for the enlisted Marine sometimes contrasted with what Marine officers said about one another, with the friction that often comes from having too many senior officers with strong personalities gathered in one command or staff.

General Conway Sums Up

During the turnover to the multinational division at Camp Babylon on 3 September, General Conway summed up the past five months in a few words: "[A]s we headed south out of Baghdad for these provinces, we did so with a certain amount of trepidation. Marines do not traditionally do nation-building or security operations. We have no doctrine for it. We were not sure where the resources would come from. And we were not sure how we would be received by the people of southern Iraq, who had seen American troops attack up through their governates. . . . [But] in some regards, a negative can become a positive. A lack of doctrine allowed us to pass some very simple rules to our Marines and soldiers. They were–treat others as you would like to be treated. Deal with the people with fairness and firmness and dignity. And among other things, we emphasized the children. They are the future of this country. It is hard to be angry with someone when he is doing good things for your children."*

*Sgt Colin Wyers, "I MEF Transfers Authority for Southern Iraq to Polish-led Division," dtd 3Sep03 (Story No. 200395144422 posted on MarineLink).

"This is the best we have had it!" laughs Master Gunnery Sergeant Paul D. Clark from Austin, the battalion's operations chief.[458]

The Marines used their imagination to defeat the enemy. They adopted a Trojan-horse approach, disguising their vehicles as Army supply trucks and then counter-ambushing the locals who had apparently been paid by Ba'athists to take potshots at the convoys. They experimented with various other ways to escort the convoys, training the Army truck drivers in convoy operations and molding Marines and soldiers into one team. Forced to seek cover some distance from the highway, the insurgents took to using increasingly sophisticated "improvised explosive devices," the small but often deadly bombs that could be rigged to detonate when vehicles passed by. The 1st Marine Division resisted advice to reply to attacks with heavy firepower, which its leaders felt might be counterproductive. Instead they decided to go to the source, that is trying to neutralize the perpetrators before they struck. This they did by collecting intelligence about them and conducting raids, often on the compounds of the relatively wealthy. It was possible to leverage the various civil affairs initiatives to get information; locals grateful for a month or two of electricity or clean water might give the Marines tips about potential insurgents, and the result might be a productive raid at dawn, a time chosen by the division to minimize the risk of harm to bystanders. Success reinforced success. When locals saw the Marines arrest and remove one perpetrator, other Iraqis were emboldened to pass on a tip that would lead to the next raid. Lieutenant Colonel Pappas commented that even if one of the attackers got away after an ambush, often the locals would approach his Marines later on and tell them where to find him.[459] Colonel Pappas' intelligence officer, Major Steven B. Manber, added that even though the task force was rich in technical collection assets, 90 percent of its successful operations stemmed from local contacts.[460]

Similar approaches were applied to more exotic locales: guarding the border with Iran in desert forts that looked like the set for a modern-day movie about the French Foreign Legion; defusing tensions in the holy Shia cities of Najaf and Karbala; uncovering and processing mass graves that were filled with the victims of the Saddam regime; patrolling the border with Saudi Arabia. Looking for weapons of mass destruction remained high on the Marines' agenda, and on that of other Coalition forces. The results were as frustrating as they were unexpected. General Conway commented that "in terms of . . . the weapons, we . . . certainly had our best guess . . . [based on what] the intelligence folks were giving us. We were simply wrong. . . . It remains a surprise to me now that we have not uncovered weapons. . . . It is not for lack of trying. We have been to virtually every ammunition supply point between the Kuwaiti border and Baghdad, and they are simply not there."[461]

By early summer, Babylon was the scene of turnover preparations. The location encompassed both the ancient city and the site of another sumptuous modern palace built by Saddam, on an artificial hill, where Marines could camp out. There being little electricity and no running water, but lots of big rooms with high ceilings and marble floors, it was a relatively cool and clean place to pitch a tent. There was a certain satisfaction to it; Marines were now in Saddam's bedroom and ballroom. Both force and division had their headquarters in Babylon, which was not far from CFLCC's forward headquarters in Bagh-

DVIC DM-SD-05-11033

Marines of Company K, 3d Battalion, 5th Marines, carrying their M16A2 rifles and gear, board a KC-130 Hercules aircraft at Blair Field in Al Kut, Iraq, for the trip home. By November 2003, all Marines and every piece of their equipment had been withdrawn.

dad. It was here that in late June they received the various parts of the Polish-led multinational division, which pitched their tents among the Marine units. There now began a process of turnovers, the first beginning in early July. Six more reliefs occurred in fairly rapid succession, not always smoothly. General Kelly reported that it was "an understatement to say that the multinational soldiers were shocked at the unexpected level of danger . . . in [their] zones," to say nothing of the Spartan living conditions and the 130-degree heat in midsummer, all of which the Marines sometimes seemed to revel in, even though there can be little doubt that many, if not most, were also ready to go home after months, and in a few cases two or three years, in theater.[462] There has always been a strange dynamic in Marine deployments. On the one hand, most Marines want very badly to march to the sound of the guns. No one wants to be left behind when the unit goes to war. But as soon as the fight is over, everyone wants to go home.

With the exception of the city of An Najaf, General Conway was able to turn over responsibility for the zone to the commander of the multinational division in a ceremony at Babylon on 3 September and send his troops south.* After the turnovers, there was a general sense of satisfaction among senior Marines with the force and division's accomplishments. Although some civilian experts noted persistent problems, particularly unemployment and inability to communicate with Iraqis, there were some undeniable statistics. For example, the number of attacks on convoys in northern Babil Province was dramatically

*Najaf was the site of a powerful car bomb attack on 29 August. This postponed the turnover, as did a number of issues that the Spanish Brigade raised. The turnover finally occurred in early October. Reflecting his general frustration with the process, one I MEF officer cracked that the turnover with the multinational division was like "stuffing cats into a seabag," while another found that many of the foreign soldiers seemed professional and ready to do their job.

DVIC DM-SD-05-11370

A multi-national delegation, including LtGen James T. Conway, second from right, and LtGen Ricardo S. Sanchez, USA, third from left, Commander, Combined and Joint Task Force 7, participate in the relief-in-place ceremony at Camp Babylon, Iraq, as Marines assigned to I Marine Expeditionary Force relinquished authority to Polish-led Coalition forces.

lower, and there were no Marines killed in action during the occupation phase, although some were wounded and, tragically, others killed in various accidents. General Mattis made the claim that since early summer the Iraqis had been "running the things that are most important to people. Are the street lights on? Is the neighborhood safe? These kinds of things are already in their hands with the Marines very much in the background."[463]

What was left for the dwindling number of Marines in Iraq and Kuwait was to complete the retrograde process. Through most of the summer, 1st Force Service Support Group Forward conducted redeployment operations from Kuwait while exercising command and control over Combat Service Support Group 11, which continued to support division. Until mid-October, I MEF Rear was still at Camp Commando, and CFLCC continued to lend its resources to support the Marines. With the assistance of the 377th Theater Support Command, Marines went through washdown and the loading onto various kinds of shipping and aircraft. The two amphibious task forces had preceded them, both setting sail in late May with heavy loads of troops and equipment, each going in a different direction. (Amphibious Task Force West had taken the more exotic route, stopping for liberty in Australia on the way.) Before flying home, other units left heavy equipment with the 1st Force Service Support Group's "Regeneration Control Element," which in turn consolidated its holdings for turn-in to the special purpose Marine air-ground task force. By 31 October, I MEF was completely gone from theater.[464]

The Marine Logistics Command morphed into the Special Purpose Marine Air-Ground Task Force under the new commanding general of 2d Force Service Support Group, Brigadier General Ronald S. Coleman, who came to Kuwait in June to replace Brigadier General Michael R. Lehnert, who was bound for a new job at Southern Command. General Coleman took up residence in the bleak expanses of Tactical Assembly Area Fox. His original charter was to repatriate and repair the Maritime Prepositioning Force equipment that had been used in Iraq, and the plan was for him to get it done by the spring of 2004. (Maritime Prepositioning Force equipment was the Marine equipment that had been prepositioned on chartered ships that made up the maritime prepositioning force.) But Coleman found a way to get the job done much faster. One of the threshold issues was whether it made more sense to do the maintenance and repairs in Kuwait and then reload the shipping, or to ship the gear first to the United States and then have it repaired at home. Various factors, including cost and temperatures so high that much of the time his Marines were literally unable to touch anything that was made of metal, made it advisable to choose the second option, and General Coleman was able to get every piece of Marine equipment out of Kuwait, and his Marines home, in time for the Marine Corps Birthday in November.[465]

General Coleman left Tactical Assembly Area Fox more or less as it had been a year earlier. Now it was almost as if the Marines had never passed through Kuwait on their way to Iraq. But there were already inklings that they would return to theater. After his work was done, General Coleman remembered a meeting with General Mattis in the summer of 2003. The division commander was concerned about the gear that was now in the Special Purpose Marine Air-Ground Task Force's charge, because he thought he would need it again shortly.[466] He seemed to know there would not be much rest for his Marines.

Notes

Chapter 1

1. Col John A. Tempone, e-mail to author, 16Jan04 (Reynolds Working Papers, Marine Corps Historical Center [MCHC], Quantico, VA).
2. See, for example, Jay E. Hines, "Operation Enduring Freedom," *21st Century Defense: U.S. Joint Command* (Clearwater, FL: Belmont International, 2002), pp. 38-46. This is an excellent overview of the operation from the CentCom point of view by that command's historian.
3. For a discussion of the historical context, see Williamson Murray and MajGen Robert H. Scales, Jr., *The Iraq War* (Cambridge, MA: Harvard University Press, 2003), pp. 48-59.
4. BGen Christian B. Cowdrey intvw, 26Apr03 (MCHC, Quantico, VA). For a snapshot of Marine operations on and after 11 September, see U.S. Marine Corps, *Operation Enduring Freedom Combat Assessment Team Summary Report* (Quantico, VA: Marine Corps Combat Development Command, 2003), pp. 47-50, hereafter MCCDC, *OEF Summary Report*. This report, by Marines assigned to MCCDC's Studies and Analysis Division, is a very useful all-around source, as are the thousands of documents and interviews that they collected.
5. Gen James L. Jones, Jr. intvw, 14Jun04 (MCHC, Quantico, VA).
6. There are numerous sources on the history of C/JTF-CM. The Cowdrey interview is a good first-hand description of the first few months of its existence. C/JTF-CM also filed an excellent set of command chronologies.
7. Task Force 58 ComdCs, 27Oct01, 26Feb02 (GRC, Quantico, VA), p. 4; hereafter TF 58 ComdC. Although it contains some idiosyncrasies, this is an unusually well-written command chronology and gives the reader an excellent picture of Marine Corps operations against the Taliban.
8. MajGen James M. Mattis, "The Professional Edge," *Marine Corps Gazette*, Feb04, p. 19; Bruce Catton, *Grant Takes Command* (Boston, MA: Little, Brown, 1969).
9. See, for example, Murray and Scales, *Iraq War*, p. 116f.
10. MCCDC, *OEF Summary Report*, p. 62.
11. TF 58 ComdC, p. 14.
12. Ibid., p. 9.
13. Ibid., p. 10.
14. Ibid., pp. 9-16.
15. Ibid.
16. LtCol Frank G. Hoffman, USMCR, "The U.S. Marine Corps in Review," *U.S. Naval Institute Proceedings*, May02, p. 84. Other articles about this operation include BGen Michael E. Ennis, "A New Operating Environment," *Marine Corps Gazette*, Aug02, p. 46f, and Capt Jay M. Holtermann, "The 15th Marine Expeditionary Unit's Seizure of Camp Rhino," *Marine Corps Gazette*, Jun02, pp. 41-43.
17. BGen James N. Mattis intvw, 20Jan02 (Naval Historical Center, Washington, DC); hereafter Mattis intvw no. 1. For background, see U.S. Marine Corps, "Expeditionary Maneuver Warfare," *Marine Corps Gazette*, Feb02, pp. A-1 through A-10. "Seabasing enables forces to move directly from ship to objectives deep inland and represents a significant advance from traditional, phased amphibious operations."
18. Col Timothy C. Wells intvw, 27Feb03 (MCHC, Quantico, VA). See also MarForPac ComdC, Jul-Dec01 (GRC, Quantico, VA).
19. MarForPac ComdC, Jul-Dec01 (GRC, Quantico, VA).
20. LtGen Earl B. Hailston intvw, 19May03 (MCHC, Quantico, VA); Wells intvw.
21. Col Nicholas E. Reynolds, "OIF Field History Journal," 2003 (MCHC, Quantico, VA), entry for 10Apr03; hereafter Reynolds, Journal.
22. TF 58 ComdC, p. 32.
23. For a discussion of this topic, see Charles J. Quilter II, *With the I Marine Expeditionary Force in Desert Shield and Desert Storm* (Washington, D.C.: History and Museums Division, 1993), pp. 8-11. See, also Jones intvw.
24. Zinni intvw.
25. Col Stephen W. Baird intvw, 31May03 (MCHC, Quantico, VA).
26. See LtCol Jeffrey Acosta, "OIF Field History Journal," 2003 (MCHC, Quantico, VA), entry for 31Mar03, hereafter Acosta, Journal. See also Wells and Hailston intvws.
27. LtGen James T. Conway intvw, 7Dec03 (MCHC,

Quantico, VA).
28. See, for example, MCCDC, *OEF Summary Report*, p. 63. Such comments were made to the author by various Marines during a visit to Tampa in December 2001.
29. MCCDC, *OEF Summary Report*, p. 57.
30. See, for example, MCCDC, *OEF Summary Report*, pp. 9-15, which argues that the operation did not fully validate expeditionary maneuver warfare but showed the utility of some of its tenets.
31. LtGen David D. McKiernan intvw, 30Jun03 (U.S. Army Center of Military History, Washington, D.C.), hereafter McKiernan intvw no. 4. To cite other examples: Gen Hailston also commented on his good relations with Gen Franks; Gen Mattis repeatedly stressed the value of establishing good relations with neighboring commanders of all Services, both joint and combined. (Hailston intvw; Mattis intvw no. 1; TF 58 ComdC) See also Reynolds, Journal, entry for 10Nov03, describing a talk by Williamson Murray and MajGen Robert H. Scales, Jr., during which they contrasted the personalities, and their relationships, in the two wars.
32. See, for example, Zinni intvw.
33. Murray and Scales, *Iraq War*, p. 47f.
34. See, for example, Bing West and MajGen Ray L. Smith, USMC (Ret.), *The March Up* (New York, NY: Putnam, 2003), p. 221, hereafter West and Smith, *March Up*. The current edition of FMFM 1 is known as MCDP 1. The original edition appeared in 1989.
35. McKiernan intvw no. 4.
36. Ibid. Franks described himself as a maverick. Tommy Franks, *American Soldier* (New York, NY: Regan Books, 2004), p. 367. For a critical appraisal of Gen Franks, see Andrew Bacevich, "A Modern Major General," *New Left Review*, Sept-Oct04, pp. 123-134.
37. Zinni intvw.
38. Acosta, Journal.
39. Murray and Scales, *Iraq War*, p. 60.
40. Kevin Peraino, "Low-Key Leader/LtGen David McKiernan Is the Soft-spoken Soldier with the Hard Job of Commanding U.S. Ground Forces in Iraq," *Newsweek* web exclusive, 19Mar03.
41. Conway intvw.
42. Murray and Scales, *Iraq War*, p. 65.
43. See, for example, Col Christopher G. Wright intvw, 4Dec03 (MCHC, Quantico, VA).
44. Reynolds, Journal, entry for 10May03.
45. Conway intvw.
46. Maj Theodore R. McKeldin, III, "OIF Field History Journal," 2003, entry for 2 April 2003; hereafter McKeldin, Journal.

Chapter 2

47. For a history of the war based largely on reporting by the *New York Times* and including coverage of political events, see, for example, Todd S. Purdum, *A Time of Our Choosing: America's War in Iraq* (New York, NY: Henry Holt, 2003), p. 20. Two other, somewhat more controversial books that cover the political background to the war are Bob Woodward, *Plan of Attack* (New York, NY: Simon and Schuster, 2004) and Richard A. Clarke, *Against All Enemies* (New York, NY: Free Press, 2004). Franks, *American Soldier*, esp. p. 315 and pp. 328-33, describes the initial taskings and briefings in late 2001.
48. Gregory Fontenot, E. J. Degen, and David Tohn, *On Point: The U.S. Army in Operation Iraqi Freedom* (Fort Leavenworth, KS: Combat Studies Institute Press, 2004), p. 31. This is a very useful and detailed early work on the conflict. On directives in 2001 to prepare for a possible war with Iraq, see, for example, Purdum, *Time of Our Choosing*, p. 20. For the point of view of a Marine planner at Headquarters Marine Corps in 2001, see, for example, LtCol James L. Western intvw, 30Dec03 (MCHC, Quantico, VA), hereafter Western intvw.
49. I MEF ComdC, Jan-Jun02 (GRC, Quantico, VA), Sec 2, p. 16.
50. Ibid., pp. 28-29, 59.
51. LtCol George W. Smith, Jr. intvw, 8Jun04 (MCHC, Quantico, VA), hereafter G.W. Smith intvw no. 2. For general background on the planning process, see also Col Christopher J. Gunther intvw, 24-25May03 (MCHC, Quantico, VA), and LtCol Gregory M. Douquet, 29Apr03 (MCHC, Quantico, VA). Although he did not become the I MEF planner until the middle of 2002, Col Gunther describes how the planning for Iraq evolved from February 2002 onward. LtCol Douquet was the senior 3d MAW planner, serving with I MEF future operations from February 2002 to April 2003.
52. Murray and Scales, *Iraq War*, p. 92.
53. Zinni intvw.
54. Murray and Scales, *Iraq War*, p. 54.
55. Ibid.
56. G. W. Smith intvw no. 2. Franks, *American Soldier*, pp. 382-431, describes the evolution of the various plans in 2002 from his point of view, especially his broad-brush briefings to the president and the secretary of defense, which led to general guidance for the planners.
57. LtCol G.W. Smith intvw, 18Mar03 (MCHC, Quantico, VA.), hereafter G. W. Smith intvw no. 1.
58. G. W. Smith intvw no. 2.

59. Reynolds, Journal, entry for 1Jul04, based on a conversation with former PP&O current operations officer Col Ronald J. Johnson.
60. G. W. Smith intvw no. 2.
61. G. W. Smith intvw no. 2; I MEF ComdC, Jan-Jul02 (GRC, Quantico, VA).
62. MCCDC, *OEF Summary Report*, p. 54; Fontenot, et al., *On Point*, p. 42.
63. McKiernan intvw no. 4.
64. Conway intvw; Hailston intvw.
65. See, for example, MajGen Robert R. Blackman intvw, 31May03 (MCHC, Quantico, VA); G. W. Smith intvw no. 2. On Gen McKiernan's approach to joint operations from a Marine point of view, see Col Marc A. Workman intvw, 30Nov02 (MCHC, Quantico, VA).
66. McKiernan intvw no. 4.
67. See, for example, the comment on this subject in Zinni intvw.
68. Fontenot, et al., *On Point*, p. 45. This conclusion assumes that the subordinate commands had the requisite clearances to access highly classified files.
69. Tom Clancy with Gen Anthony Zinni (Ret) and Tony Koltz, *Battle Ready* (New York, NY: G. P. Putnam's Sons, 2004), p. 315.
70. Zinni intvw.
71.MCCDC, Operation *Iraqi Freedom Combat Assessment Team Summary Report*, spring 2004 (Studies and Analysis Division, Working Draft), p. 40, hereafter MCCDC, *OIF Summary Report*.
72. There was a complicated set of agreements about this, which are examined in detail in LtCol LeRoy D. Stearns, *The 3d Marine Aircraft Wing in Desert Shield and Desert Storm* (Washington, D.C.: Marine Corps History and Museums Division, 1999), pp. 45-49, hereafter Stearns, *3d Marine Aircraft Wing*. See also Michael R. Gordon and Gen Bernard E. Trainor, *The Generals' War* (Boston, MA: Little, Brown, 1995), p. 311. This is a basic source on Desert Shield/Desert Storm.
73. MajGen James F. Amos intvw, 16May03 (MCHC, Quantico, VA); Hailston intvw.
74. Stearns, *3d Marine Aircraft Wing*, p. 47.
75. See G. W. Smith intvw no. 2; Reynolds, Journal, entry for 29Jul03, describing a postwar briefing by the general officers who commanded I MEF's major subordinate commands. Capt Ryan M. Walker intvw, 22Apr03 (MCHC, Quantico, VA). By and large, once the war started, the system worked. See MCCDC, *OIF Summary Report*, p. 40.
76. Fontenot, et al., *On Point*, p 13.
77. Murray and Scales, *Iraq War*, p. 154-183. For a comment on Iraqi exiles in general, see, for example, Clancy, *Battle Ready*, p. 345.
78. G. W. Smith intvw no. 2.
79. The issue of synchronicity is addressed in, among other sources, LtCol Richard T. Johnson intvw, 26Apr03 (MCHC, Quantico, VA); McKeldin Journal, entry for 28Feb03; and LtCol Paul J. Kennedy intvw, 6Nov03 (MCHC, Quantico, VA). See, for example, G. W. Smith intvw no. 2, and Douquet intvw.
80. There are numerous sources on this topic. G. W. Smith intvw no. 2 contains details of meetings on the subject, including a meeting between Gens McKiernan and Conway on 29 January 2003 when they discussed, and agreed on, this subject.
81. Quoted in House of Commons Defence Committee, *Lessons of Iraq* (London, UK: Stationery Office, 2004), vol. 1, pp. 34-35. This is a useful source with many lengthy, and candid, quotations by British officers. It appears to be more candid than the Ministry of Defence's "lessons learned" publications on the war.
82. Fontenot, et al., *On Point*, pp. 93-94; Terry Moran intvw, 23Aug03 (U.S. Army Center of Military History, Washington, D.C.); Col Reed R. Bonadonna, "Notes from Address by LtGen Conway, CG I MEF to Officers of Task Force Tarawa," 1Mar03 (Copy in Reynolds Working Papers, MCHC, Quantico, VA); I MEF ComdC, Jan-Jun03 (GRC, Quantico, VA), "FutOps" (Documents), "I MEF ROC Drill, 10Mar03."
83. G. W. Smith intvw no. 2; see also Gunther intvw.
84. Fontenot, et al., *On Point*, p. 74.
85. G. W. Smith intvw no. 1.
86. Maj Evan A. Huelfer intvw, 16Mar03 (U.S. Army Center of Military History, Washington, D.C.); G. W. Smith intvw no 2; Fontenot, et al., *On Point*, p. 45.
87. On the neoconservative movement, see James Mann, *Rise of the Vulcans: The History of Bush's War Cabinet* (New York, NY: Viking, 2004). On some of the neoconservative plans that were presented to Franks, see Franks, *American Soldier*, p. 373. For a general source on the planning process, see Woodward, *Plan of Attack*. Although Woodward and Franks cover much of the same ground, Woodward describes a slightly different dynamic from Franks, suggesting that the Pentagon, and not CentCom, generally took the lead in proposing various plans. For a more critical view of the development of the administration's policy on Iraq, see W. Patrick Lang, "Drinking the Kool-Aid," *Middle East Policy*, summer 2004, pp. 39-60.
88. Although Franks said he had a productive relationship with Donald Rumsfeld and liked working with the Secretary of Defense, he also expressed

reservations about his "centralized management style." See Franks, *American Soldier*, pp. 313, 333, 362, 373, and 545. Col Kevin Benson, the senior CFLCC planner, reported similar frustrations with the Pentagon during the planning stage. Colonel Kevin Benson, "Brief at Naval War College," 25Aug04 (Copy in Reynolds Working Papers, MCHC, Quantico, VA). See also Anthony H. Cordesman, *The Iraq War: Strategy, Tactics, and Military Lessons* (Washington, D.C.: CSIS, 2003), p. 150, and questions about "alleged meddling . . . by the Office of the Secretary of Defense—especially in the planning process" down to the CFLCC/MEF level, in U.S. Naval Institute, "Interview: LtGen James T. Conway, USMC," *U.S. Naval Institute Proceedings* (November 2003). Some have suggested that war planning for Iraq needs to be understood in the context of relations between the secretary and the military starting in 2000. See, for example, Elaine M. Grossman, "To Understand Insurgency in Iraq: Read Something Old, Something New," *Insider*, 2Dec04, containing a comment by LtGen Gregory Newbold, who retired in 2002 from a senior position in the J-3 at the Pentagon, that a cautionary tale about the Pentagon during the Vietnam years "contains remarkable parallels with today's environment." The book is H. R. McMaster, *Dereliction of Duty* (New York, NY: Harper Collins, 1997). It will be a number of years before the sources are available for anyone to analyze the Iraq War in the same way.

89. Zinni intvw; Jones intvw.

90. See, for example, LtCol Ronald J. Brown, *With Marines in Operation Provide Comfort: Humanitarian Operations in Northern Iraq* (Washington, D.C.: History and Museums Division, 1995); Gordon and Trainor, *The Generals' War*, p. 459.

91. G. W. Smith intvw no. 2; Huelfer intvw; Moran intvw. Moran, a retired Army lieutenant colonel, was General McKiernan's special assistant, privy to a great deal of information about the plan for the war.

92. Huelfer intvw; Benson, "Brief at Naval War College." Benson and Huelfer cover much the same ground.

93. Huelfer intvw.

94. G. W. Smith intvw no. 2.

95. There is an excellent discussion of this topic in Fontenot, et al., *On Point*, pp. 44-47. Once again, Huelfer and Smith also offer excellent descriptions of the process from the planner's perspective. Huelfer intvw; G. W. Smith intvws nos. 1 and 2.

96. Huelfer intvw; G. W. Smith intvw no. 2.

97. Huelfer intvw.

98. G. W. Smith intvw no. 2.

99. G. W. Smith intvw no. 2.

100. G. W. Smith intvw no. 2; intvw with LtCol Yanni Marok, Royal Marines, 14Apr03 (MCHC, Quantico, VA). Marok joined I MEF future operations staff in September and developed an excellent relationship with other planners, including Smith. His interview is particularly useful for the data it contains on Al Faw.

101. Marok intvw. This occurred in the 12-13 December time frame at a conference in Bahrain. Col Gunther was acting with the knowledge and approval of his superiors.

102. G. W. Smith intvw no. 2: Franks, *American Soldier*, p. 396, contains his schematic of the various lines of operation.

103. I MEF ComdC, Jul-Dec02 (GRC, Quantico, VA); Col George F. Milburn intvw, 1Mar03 (MCHC, Quantico, VA); P. J. Kennedy intvw.

104. Reynolds Journal, entry for 19Jul04, quoting from journal kept by one of the officers present.

105. Reynolds Journal, entry for 18Jul04.

106. LtGen David D. McKiernan intvw, 20Jun03 (U.S. Army Center of Military History, Washington, D.C.), hereafter McKiernan intvw no. 3. See also Blackman intvw, Workman intvw, and the television documentary about CFLCC by CNN Presents, "Inside the War Room" (Atlanta, GA: CNN, 2003), hereafter CNN, "The War Room."

107. There are numerous sources about this approach, defined as "a methodology for planning, executing, and assessing operations . . . to attain the effects required to achieve desired . . . outcomes." Edward C. Mann III, Gary Endersby, Thomas R. Searle, *Thinking Effects: Effects-Based Methodology for Joint Operations* (Maxwell Air Force Base, AL: Air University Press, 2002), p. 2.

108. Col Gregory J. Plush intvw, 15Apr03 (MCHC, Quantico, VA); Workman intvw.

109. Fontenot, et al., *On Point*, p. 43; Blackman intvw; CNN, "The War Room."

110. Col Patrick J. Burger intvw, 16Apr03 (MCHC, Quantico, VA). There were approximately 12 Marine liaison officers at CFLCC.

111. See MCCDC, *OEF Summary Report*, p. 11, reporting its finding that providing adequate numbers of staff officers for such components was a challenge for the Marine Corps.

112. Blackman intvw. See also Workman and Cowdrey intvws. BGen Cowdrey was one of the "January" augmentees. In February 2003 there were some 1,300 members of the CFFLC staff. LtCol Terrance J. Johns and Maj Robert F. McTague II intvw, 22Feb03 (CFLCC Military History Group intvw, 322-332).

113. Huelfer intvw.
114. See, for example, Fontenot, et al., *On Point*, p. 41.
115. Jones intvw.
116. Blackman intvw. This tracks with the comments by Gen McKiernan and Moran with respect to the plans that were under discussion in the fall of 2003, but diverges somewhat from two of the lead planners' comments. G. W. Smith intvw no. 2; R. T. Johnson intvw; I MEF ComdC, Jul-Aug02 (GRC, Quantico, VA), see 2 narrative summaries for G-3 Future Operations and G-5 Plans. For a more general discussion of planning issues at this time, see Cordesman, *Iraq War*, pp. 149-159.
117. LtGen David D. McKiernan intvw, 19Dec02 (U.S. Army Center of Military History, Washington, DC), hereafter McKiernan intvw no. 1. See also R. T. Johnson intvw.
118. Intvw with LtGen David D. McKiernan, 1May03 (Copy in Reynolds Working Papers (MCHC, Quantico, VA), hereafter McKiernan intvw no. 2. This was a group interview with officers representing both the Military History Group at Camp Doha and various Army lessons-learned initiatives. Much of the same ground is also covered in McKiernan intvw no. 1.
119. Huelfer intvw.
120. Moran intvw.
121. I MEF ComdC, Jan-Jun03 (GRC, Quantico, VA), "command element," p. 41; Blackman intvw.
122. Reynolds, Journal, entry for 1Jul04.
123. Zinni intvw.
124. I MEF ComdC, 1Jan03-30Jun03 (GRC, Quantico, VA), Future Operations Folder, "Chg 1 to I MEF OPLAN 1003V," 16Mar03.
125. McKiernan intvw no. 4; G. W. Smith intvw no. 2.
126. G. W. Smith intvw no. 2.
127. Ibid.
128. BGen Richard F. Natonski intvw, 26Mar04 (MCHC, Quantico, VA), hereafter Natonski intvw no. 2.
129. See, for example, Huelfer intvw.
130. McKiernan intvw no. 1.
131. Ibid. The phases are nicely laid out in Fontenot, et al., *On Point*, p. xxiii.
132. G. W. Smith intvw no. 1; P. J. Kennedy intvw. The Conway quote is from a 7 August 2003 interview published in the *U.S. Naval Institute Proceedings* (November 2003), which also appeared on the internet at www.usni.org/proceedings/articles03. See also Reynolds Journal, entry for 1 July 2004.
133. LtCol Jim Hutton intvw, Royal Marines, 4May03 (MCHC, Quantico, VA).
134. Col George F. Milburn intvw, 3Aug03 (MCHC, Quantico, VA); hereafter Milburn intvw no. 2. This is another great topic for a paper, if not a dissertation. The researcher could compare Marine, Army, and British plans and orders in this conflict and consider questions like: Had the British adapted to the information age, or were they simply applying lessons long since learned? How has American planning adapted to the information age? In a war of "shock and awe," when do planners reach the point of diminishing returns?
135. Rick Atkinson, *In the Company of Soldiers* (New York, NY: Henry Holt, 2004), p. 26. This is, in my view, one of the best books about the war by an embedded journalist, or by anyone else. Atkinson does an excellent job of portraying the frame of mind of the staff of the 101st Airborne Division and of putting its accomplishments in context.
136. Bonadonna, "Notes from Address by LtGen Conway."
137. McKiernan intvw no. 4.
138. Huelfer intvw.
139. See, for example, Atkinson, *In the Company of Soldiers*, p. 107.
140. McKiernan intvw no. 2, 1May03 (group interview at Camp Doha, KU, copy at MCHC, Washington, DC). Fontenot, et al., *On Point*, pp. 99,102, offers a summary of the enemy order of battle as it appeared before the campaign. See also I MEF ComdC, Jan-Jun03 (GRC, Quantico, VA), especially Future Operations folder containing "Chg 1 to MEF OPLAN 1003V," dated 16Mar03, and containing an assessment of enemy capabilities.
141. Reynolds, Journal, entry for 1Jul04.
142. MajGen James N. Mattis intvw, 23Jan04 (MCHC, Quantico, VA); Wright intvw.
143. Bonadonna, "Notes from Address by LtGen Conway."
144. Steve Col, "Hussein Was Sure of Own Survival: Aide Says Confusion Reigned on Eve of War," *The Washington Post*, 3 Nov03, p. A-1. This would certainly not have been out of character for Saddam Hussein, many of whose past decisions were not thought through, to put it mildly. See also Franks, *American Soldier*, p. 558, for a comment on the work of a Joint Forces Command lessons-learned team that debriefed Iraqi officers after the war. The team found that while some Iraqi units were well prepared, Iraqi command and control, especially at senior levels, was next to nonexistent. Finally, see various sources on a CIA report made public in 2004, including Douglas Jehl, "Inspector's Report Says Hussein Expected Guerrilla War," *The New York*

Times, 8Oct04, p. A-6.
145. These issues surfaced repeatedly in oral history interviews. See for example, Blackman intvw; Reynolds, Journal, entry for 1Jul04; G. W. Smith intvw no. 2; Zinni intvw; and Clancy, *Battle Ready*, pp. 19-20.

Chapter 3

146. Conway intvw.
147. Capt Lara A. Bennett, et al., *No Better Friend, No Worse Enemy* (Camp Pendleton, CA: 1st Marine Division, 2004), pp. 6-8. This book has pictures and text about the division in Operation Iraqi Freedom and was distributed by the Marine Corps Association.
148. Wright intvw; Col John A. Toolan intvw, 10Dec03 (MCHC, Quantico, VA).
149. 1stMarDiv ComdC, Jan-Jul03 (GRC, Quantico, VA), sec 2, chap 3, p. 21. There are numerous other sources on the division's preparations for war, including Toolan intvw; Wright intvw; and Col Joseph F. Dunford intvw, 2May03 (MCHC, Quantico, VA).
150. Jones intvw.
151. BGen Edward G. Usher III intvw, 19Mar03 (MCHC, Quantico, VA), hereafter Usher intvw no. 1; the Commanders and Staff of 1st FSSG, "Brute Force Logistics," *Marine Corps Gazette*, Aug03, pp. 34-39. A field historian was attached to 1st Force Service Support Group during Operation Iraqi Freedom. She was Maj Melissa D. Kuo, who is preparing a monograph on combat service support in OIF I. I am indebted to her for reviewing my remarks on logistics and making suggestions.
152. Hailston intvw; Usher intvw no. 1; LtCol Michael. D. Visconage, "OIF Field History Journal," 2003 (MCHC, Quantico, VA), entry for 15Mar03, hereafter Visconage, Journal; BGen Michael R. Lehnert intvw, 8May03 (MCHC, Quantico, VA). One of the Marine field historians assigned to cover OIF I, CWO-2 William E. Hutson, conducted numerous interviews at the Marine Logistics Command that are available at the MCHC.
153. Reynolds, Journal, entry for 29Jul03. This entry contains detailed notes on a set of briefings given by the OIF commanders, including Generals Conway, Mattis, and Amos, at Quantico. The author worked for General Amos at The Basic School.
154. This is a topic that the U.S. Air Force has explored over the years. See, for example, Robert P. Givens, *Turning the Vertical Flank* (Maxwell Air Force Base, AL: Air University Press, 2002). Col Givens outlines the criteria for thinking of air as a maneuver force, able to exert influence over enemy units and terrain on its own.
155. These topics are addressed in two excellent interviews with the 3d MAW G-3, Col Jonathan G. Miclot, on 20Mar03-19Apr03 (MCHC, Quantico, VA).
156. 3d MAW ComdC, Jan-Jul03 (GRC, Quantico, VA), sec 2, p. 6.
157. See ibid., and, for a discussion of FARPs, LtCol David P. Lobik intvw, 24Apr03 (MCHC, Quantico, VA).
158. 3d MAW ComdC, Jan-Jul03 (GRC, Quantico, VA), sec 2, "G-5 Plans and Exercises."
159. Reynolds, Journal, entry for 29Jul03.
160. MCCDC, *OIF Summary Report*, p. 18. This is a quote from a video teleconference; members of the Combat Assessment Team routinely attended these VTCs and made shorthand notes.
161. R. T. Johnson intvw; I MEF ComdC, Jan-Jun03 (GRC, Quantico, VA), sec 2, p. 6.
162. See, for example, MCCDC, *OIF Summary Report*, pp. 20-21.
163. LtCol James W. Western intvw, 20Dec03 (MCHC, Quantico, VA).
164. Peraino, "Low-Key Leader." There are other representative discussions of this issue in Mattis intvw no. 1; Lehnert intvw; and Moran intvw.
165. 2d MEB ComdC, Jan-Jun03 (GRC, Quantico, VA), sec 3, "Significant Events."
166. 1stMarDiv ComdC, Jan-Jun03 (GRC Quantico, VA), sec 2, chap 2, p. 28.
167. The offload took about 16 days, a few days less than the allotted time. MCCDC, *OIF Summary Report*, p. 55.
168. Ibid., p. 20. See also Milburn intvw no. 2 for comments by the I MEF G-5 on the process.
169. For a discussion of this topic, see Plush intvw.
170. Maj Grant A. Williams, "A Marine's Eye View of Kuwait," Milinet (an internet service provider that posted news about the military on the web), 27Feb03. This excerpt is from a report originally dated 19Feb03 (Copy in Reynolds Working Papers, MCHC, Quantico, VA).
171. Peter Baker, "In War, Plans Yield to Improvisation," *The Washington Post*, 24Mar03, p. A-16.
172. Ibid.
173. Col Steven A. Hummer intvw, 13Feb04 (MCHC, Quantico, VA). See also McKeldin Journal, entry for 7Feb03.
174. BGen Cornell A. Wilson intvws, 6Mar03, 3Apr03, 17May03 (MCHC, Quantico, VA). MCHC holds a number of other interviews on Consequence Management, along with the task force's command chronology.

175. Reynolds, Journal, entry for 8Apr03.
176. Gen James N. Mattis, "Memorandum for All Commanders," 20Dec02 (Copy in Reynolds Working Papers, MCHC, Quantico, VA).
177. McKeldin, Journal, entry for 7Feb03.
178. For additional details see, I MEF ComdC, Jan-Jun03 (GRC, Quantico, VA), sec 2, "FutOps Narrative Summary"; McKeldin, Journal; CWO-2 William E. Hutson, "OIF Field History Journal," (MCHC, Quantico, VA), entry for 4 Feb. 2003, hereafter Hutson, Journal. There is a discussion of this phase in the intvw of the I MEF's current operations officer, who pointed out all of the work involved in bringing the I MEF staff together and producing the plan. Col Dennis Judge intvws, 11Aug03 and 4Nov03 (MCHC, Quantico, VA), hereafter Judge intvws no. 1 and no. 2.
179. I MEF ComdC, Jan-Jun03 (GRC, Quantico, VA), Future Operations Folder, "I MEF ROC Drill, 10 Mar 03"; McKeldin, Journal, entries for 27Feb03 and 10Mar03. For Task Force Tarawa, see Col Reed R. Bonadonna, "Field Historian Journal" (MCHC, Quantico, VA), entries for 2 and 10Mar03, hereafter Bonadonna, Journal. This journal is very well written and gives the reader a good sense of what it was like to participate in OIF as a member of Task Force Tarawa. I recommend it highly both to the general reader and the reader who wants to learn more about Task Force Tarawa. In the run-up to combat operations, subordinate commanders down to the division level would use the terrain model at Doha again to brief General McKiernan and his staff on their plans. MajGen James D. Thurman intvw, 27May03 (Copy in Reynolds Working Papers, MCHC, Quantico, VA).
180. Mattis intvw no. 1; Hutson Journal, entry for 27Feb03. See also 1stMarDiv ComdC, Jan-Jun03 (GRC, Quantico, VA), sec 2, "Narrative Summary," chap 3. The division has written an unusually comprehensive narrative summary from its point of view.
181. Bonadonna, Journal, entry for 1Mar03; Conway intvw.
182. Bonadonna, "Notes from Address by LtGen Conway."
183. Mark Mazetti, "Fighting Words: A Marine Commander Readies His Troops for Combat," *U.S. News and World Report* web exclusive, 17Mar03; McKeldin Journal, entry for 16Mar03.
184. Conway intvw; Murray and Scales, *Iraq War*, p. 65.
185. Conway intvw; Nick Parker, "Countdown to Conflict," *The Mirror*, 15Mar03, pp. 6-7; Vanessa Allen, "'Great to Have You Aboard' U.S. General Tells Brits," The Press Association (web-based wire service), 14Mar03.
186. Hutson, Journal, entry for 27Feb03.
187. Bonadonna, Journal, entry for 27Feb03.
188. Bennett, et al., *No Better Friend*, p. 5.
189. Mattis intvw no. 1.
190. Visconage, Journal, entry for 4Mar03.
191. MajGen James F. Amos, e-mail to multiple addressees, "Message from the Wing CG," 20Mar03 (Col Charles J. Quilter II CD-ROM, Reynolds Working Papers, MCHC, Quantico, VA). See, for example, Bonadonna Journal, and Maj Melissa D. Kuo, "Field History Journal" (MCHC, Quantico, VA), entry for 18Mar03; hereafter Kuo, Journal.
192. Sarah Oliver, "Not the Beginning, It's the End–The Words of LtCol Tim Collins," *Mail on Sunday*, 16Mar03, pp. 6-7.
193. Representative of other articles in the British press about the affair are Jack Grimston and John Elliott, "First Picture of Man Who Denounced Tim Collins," *Sunday Times* (London), 25Ma03, p. 1; and "Col Tim: The Making of a Modern Hero," *Independent on Sunday* (London), 25May03, Features Section. Even though somewhat condescending to Americans and reservists, the last article is a thoughtful analysis of Collins' participation in the war, from his prewar speech through the investigation.
194. "The Smearing of Col Tim; The Eve of Battle Speech," *Daily Mail* (London), 22May03, p. 9.
195. See, for example, McKeldin, Journal, entries for 5-6Mar03, and Reynolds, Journal, entry for 1 July 2004, reporting on a talk with the G-3 of Task Force Tarawa.
196. McKeldin, Journal, entry for 18Mar03.

Chapter 4

197. McKiernan intvw no. 1.
198. Bonadonna, "Notes from Address by LtGen Conway."
199. Ibid.
200. I MEF Sitrep 181800Z to 191759ZMar03 (Copy in Reynolds Working Papers, MCHC, Quantico, VA).
201. Fontenot, et al., *On Point*, p. 93; I MEF Sitrep 171800Z to 181759ZMar03 (Copy in Reynolds Working Papers, MCHC, Quantico, VA); McKeldin Journal, entry for 18 March 2003; I MEF ComdC, Jan-Jun03 (GRC, Quantico, VA), command element, sec 2, Current Operations Narrative.
202. See, for examples, Maj Phillip N. Frietze intvw, 14May03 (MCHC, Quantico, VA), and Maj Daniel E. Longwell intvw, 1May03 (MCHC, Quantico, VA); I MEF ComdC, Jan-Jun03 (GRC, Quantico, VA), Future

Operations, sec 2.
203. I MEF Sitrep 181800Z to 1981759ZMar03 (Copy in Reynolds Working Papers, MCHC, Quantico, VA).
204. I MEF Sitrep 191800Z to 201759ZMar03 (Copy in Reynolds Working Papers, MCHC, Quantico, VA).
205. See, for example, Kuo, Journal, entry for 20Mar03.
206. Col John C. Coleman intvw, 11Dec03 (MCHC, Quantico, VA).
207. McKeldin, Journal, entry for 20Mar03; Kuo, Journal, entry for 20Mar03. Another description of the day can be found in Peter Baker, "Overtaken by Events, the Battle Plans Are Tossed Aside," *The Washington Post*, 21Mar03, p. A-20.
208. I MEF Sitrep 191800Z to 201759ZMar03 (Copy in Reynolds Working Papers, MCHC, Quantico, VA); McKeldin, Journal, entry for 20Mar03.
209. Dunford intvw.
210. I MEF Sitrep 191800Z to 201759ZMar03 (Copy in Reynolds Working Papers, MCHC, Quantico, VA); General Mattis' comment is in the 20-21Mar03 Sitrep.
211. I MEF Sitrep 181800ZMar03 to 191759Mar03 (Copy in Reynolds Working Papers at MCHC, Quantico, VA). On the issue of synchronicity, see 1stMarDiv ComdC, Jan-Jun03 (GRC, Quantico, VA), sec 2, chap 2, p. 12, and chap 3, p. 6.
212. 3d MAW ComdC, Jan-Jun03, sec 3, "Sequential Listing of Significant Events." See also Acosta, Journal, entry for 20May03, containing an excellent overview of the war from the standpoint of the I MEF G-3, Col Larry Brown.
213. See, for example, 3d MAW ComdC, Jan-Jun03 (GRC, Quantico, VA), sec 2, p. 6; Lobik intvw.
214. GySgt Melba L. Garza intvw, 26May03 (MCHC, Quantico, VA).
215. LtCol Gary Smythe, USA, "1st Marine Division Fire Support Coordination during Operation Iraqi Freedom," *Marine Corps Gazette*, Jun04, p. 31.
216. I MEF Sitrep 201800Z to 211759ZMar03 (Copy in Reynolds Working Papers, MCHC, Quantico, VA).
217. I MEF sitreps 201800Z20Mar03 to 211759ZMar03, 211800ZMar03 to 221759ZMar03, and 221800ZMar03 to 231759ZMar03 (Copies in Reynolds Working Papers, MCHC, Quantico, VA). See also 1stMarDiv ComdC, Jan-Jun03 (GRC, Quantico, VA), sec 2, chap 4.

Chapter 5

218. Christopher Cooper, "How a Marine Lost His Command in the Race to Baghdad," *Wall Street Journal* Online.
219. Capt Daniel J. Wittnam intvw, 1May03 (MCHC, Quantico, VA); Bonadonna, Journal, entries for 2 and 10Mar03; 2d MEB ComdC, Jan-Jun03 (GRC, Quantico, VA), sec 2.
220. Maj Daniel T. Canfield intvw, 12Dec03; LtCol Rickey L. Grabowski intvw, 6Apr03; LtCol Royal P. Mortenson intvw, 7Nov03; Col Ronald L. Bailey intvw, 8May03, hereafter Bailey intvw (MCHC, Quantico, VA).
221. I MEF Sitrep 211800ZMar03 to 221759ZMar03 (Copy in Reynolds Working Papers, MCHC, Quantico, VA), p. 125.
222. I MEF Sitrep 221800ZMar03 to 231759ZMar03 (Copy in Reynolds Working Papers, MCHC, Quantico, VA).
223. See West and Smith, *March Up*, pp. 41, 59; Company Commanders, 1st Battalion, 2d Marines, "The Battle of An Nasiriyah," *Marine Corps Gazette*, Sep03, pp. 40, 46, hereafter Company Commanders, "Battle of An Nasiriyah." This is a good general source on the battle.
224. Bailey intvw.
225. See 2d MEB ComdC, Jan-Jun03 (GRC, Quantico, VA), sec 2.
226. Natonski intvw no. 2; BGen Richard F. Natonski intvw, 25Apr03 (MCHC, Quantico, VA), hereafter Natonski intvw no. 1; Bailey intvw.
227. See Maj Walker M. Field, "Marine Artillery in the Battle of An Nasiriyah," *Marine Corps Gazette*, Jun04, pp. 26, 30; Natonski intvw no. 2. The Bonadonna journal entries for 31 March 2003 and 30 April 2003 discuss his conclusion that the "big problems" were scant intelligence and fire support. See also Gary Livingston, *An Nasiriyah: The Fight for the Bridges* (Vernon, NY: Caisson Press, 2004), pp. 172-173, hereafter Livingston, *An Nasiriyah*. This is a useful book with a number of oral histories.
228. Bailey intvw. Col Bailey elaborated on this interview in a conversation with Col Bonadonna, which was subsequently relayed to the author.
229. Livingston, *An Nasiriyah*, p. 70.
230. Company Commanders, "Battle of An Nasiriyah"; Bonadonna Journal, entry for 30Apr03.
231. SSgt Lonnie O. Parker intvw, 29Mar03 (MCHC, Quantico, VA).
232. Livingston, An Nasiriyah, p. 91; Wittnam intvw; Natonski intvw no. 2; Parker intvw.
233. 1stLt Michael S. Seely intvw, 3May03 (MCHC, Quantico, VA).
234. Wittnam intvw.
235. Ibid.
236. Livingston, An Nasiriyah, p. 101.
237. Parker intvw.
238. Col Reed R. Bonadonna, "A Short History of

Task Force Tarawa," unpublished manuscript included with Task Force Tarawa ComdC, Jan-Jun03 (GRC, Quantico, VA).
239. Natonski intvw no. 2. See, for example, "Interview: Lieutenant General James T. Conway, USMC," *U.S. Naval Institute Proceedings* (Nov 2003).
240. I MEF Sitrep 241800ZMar03 to 25175ZMar03 (Copy in Reynolds Working Papers, MCHC, Quantico, VA).
241. See, for example, Reynolds, Journal, entries for 1Apr03, 3Apr03; Cordesman, *Iraq War*, p. 390, quoting a report: "Nasiriyah saw the culmination of this confusion over MOUT [military operations in urban terrain]. Commanders were not prepared to go in and clear a town no bigger than Victorville"; Helene Cooper, "U.S. Troops Bypass Cities to Avoid Urban Warfare," *Wall Street Journal*, 28Mar03; Bonadonna, Journal, entries for 28Mar03, 31Mar03.
242. Natonski intvw no. 2; Maj [first name unknown] Bierman, "Summary of Comments Made by Col Johnson, G-3 Task Force Tarawa," as posted on CFLCC J-2 website 3Apr03 (Copy in Reynolds Working Papers, MCHC, Quantico, VA).
243. McKiernan intvw no. 4. See also West and Smith, *March Up*, pp. 35, 41, 48, and C. Cooper, "How a Marine Lost His Command." Cooper discusses Dowdy's thought process.
244. Natonski intvw no. 2; Reynolds Journal, entry for 1Apr03; Wittnam intvw; CentCom, "News Release 04-03-51–A-10 Friendly Fire Investigation Completed," 29Mar03, hereafter CentCom News Release. See also Hector Becerra, Robert J. Lopez, and Rich Connell, "Report Details 'Friendly Fire' Casualties in Deadly Battle," *Los Angeles Times*, 28Mar03. The report itself was placed online by Globalsecurity.org. Its title is M.S. Central Command, "Investigation of Suspected Friendly Fire Incident Near An Nasiriyah, Iraq, 23 March 03," dtd 6Mar03. For a thoughtful overview of friendly fire incidents during OIF-I, see Peter Pae, "'Friendly Fire' Still a Problem," *Los Angeles Times*, 16May03.
245. Bonadonna, "Short History"; Bonadonna, Journal, entry for 23-24Mar03; West and Smith, *March Up*, pp. 47, 52.
246. CentCom News Release.
247. West and Smith, *March Up*, p. 47; LtCol Willard A. Buhl intvw, 4Nov03 (MCHC, Quantico, VA).
248. Conway intvw; Buhl intvw; Natonski intvw no. 2; West and Smith, *March Up*, p. 45.
249. Bonadonna, Journal, entry for 23Mar03. For additional impressions of an officer in theater, see Kuo, Journal, entries for 23-24Mar03. For a relatively mild example of press commentary, see Peter Baker, "In War, Plans Lead to Improvisation," *The Washington Post*, 24Mar03, p. A-16.
250. McKiernan intvw no. 4.
251. McKeldin, Journal, entry for 23Mar03.
252. McKiernan intvw no. 4.
253. Quoted in Bennett, et al., *No Better Friend*, p. 51. What appears to be the original quote appears in Peter Baker, "Arab Volunteers Draw U.S. Scrutiny; Marine Commander Outraged by Willingness to 'Hide Behind Women'," *The Washington Post*, 9Apr03, p. A-31.
254. Amos intvw.
255. Mattis intvw no. 1; I MEF Sitrep 241800ZMar03 to 251759ZMar03 (Copy in Reynolds Working Papers, MCHC, Quantico, VA).
256. BGen Edward G. Usher III intvw, 11May03 (MCHC, Quantico, VA), hereafter Usher intvw no. 2.
257. 1stMarDiv ComdC, Jan-Jun03, sec 2, chap 5.
258. Bonadonna, Journal, entry for 25Mar03.
259. Hummer intvw.
260. Judge intvw no. 1.
261. Reynolds, Journal, entry for 26Mar03; I MEF Sitrep 261800ZMar03 to 271759ZMar03 (Copy in Reynolds Working Papers, MCHC, Quantico, VA), which contains a reference to the suspension of air operations at the Jalibah airfield until 0300Z on that day. Atkinson, *In the Company of Soldiers*, reports that the skies finally cleared on the morning of 27 March.
262. Atkinson, *In the Company of Soldiers*, pp. 171-172.
263. McKiernan intvw no. 3.

Chapter 6

264. Text of Navy Cross citation, approximate date March 2004. See Cpl Jeremy Vought, "Rochester, N.Y., Marine Receives Navy Cross," 6May04. Available at http://www.usmc.mil/marinelink/mcn2000.nsf/lookupstoryref/200456162723. See also 1stLt Brian R. Chontosh intvw, 4May03 (MCHC, Quantico, VA).
265. Mattis intvw no. 1.
266. 1stMarDiv ComdC, Jan-Jun03 (GRC, Quantico, VA), sec 2, executive summary, p. 2.
267. Kuo, Journal, entry for 24Mar03.
268. Usher intvw no. 2; I MEF Sitrep 241800ZMar03 to 251759ZMar03 (Copy in Reynolds Working Papers, MCHC, Quantico, VA).
269. BGen Michael R. Lehnert and Col John E. Wissler, "Marine Logistics Command, Sustaining Tempo on the 21st Century Battlefield," *Marine Corps Gazette*, Aug03, pp. 30, 33.

270. BGen Michael R. Lehnert, 7May03 (MCHC, Quantico, VA). The division G-3 commented on shortages from his perspective in interview with LtCol Clarke R. Lethin, 6Nov03 (MCHC, Quantico, VA). For a frank discussion of the MLC and praise for General Lehnert's innovative leadership, see Hutson, Journal, entry for 25Apr03.
271. Usher intvw no. 2.
272. Commanders and Staff of 1st FSSG, "Brute Force Combat Service Support: 1st Force Service Support Group in Operation Iraqi Freedom," *Marine Corps Gazette*, Aug03, p. 37.
273. Mattis intvw no. 1. The stop order is reported in I MEF Sitrep 261800ZMar03 to 271759ZMar03 (Copy in Reynolds Working Papers, MCHC, Quantico, VA).
274. See, for example, comments by the I MEF G-3, Larry K. Brown, in Acosta, Journal, entry for 20May03 (describing a postwar symposium in Bahrain and Brown's comment to the effect that the Army had wanted a 30-day pause), and G. W. Smith intvw no. 2. Atkinson, *In the Company of Soldiers*, pp. 168, 171, 177, offers a good reflection of thinking by some senior army officers. They had a combination of concerns over supply shortages and the *Fedayeen*, leading to a natural inclination to wait on reinforcements.
275. I MEF ComdC, Jan-Jun03 (GRC, Quantico, VA), future operations folder, "CG Guidance/discussion with OPT 25 March 03." See also Elaine Grossman, "Marine General: Iraq War Pause 'Could Not Have Come at Worse Time'," *Inside the Pentagon*, 2Oct03; Conway intvw.
276. Fontenot, et al., *On Point*, p. 245, relies on two basic kinds of sources, Army note takers who were present at the meeting, and his interview with General McKiernan on 8Dec03. The note takers' records are the only firsthand, contemporary, and currently available, source on the meeting.
277. Quoted in Fontenot, et al., *On Point*, p. 245; see also Conway intvw.
278. McKeldin, Journal, entry for 28Mar03. The source for the "several days" quote is "CFLCC Update, 26 Mar 03" (OIF CD No. 65, GRC, Quantico, VA).
279. I MEF Sitrep 271800ZMar03 to 281759ZMar03 (Copy in Reynolds Working Papers, MCHC, Quantico, VA).
280. McKeldin, Journal, entry for 29Mar03.
281. "Lt Gen J Conway Holds Defense News Briefing via Teleconference from Iraq," Federal Document Clearing House Political Transcript of Interview, 30May03, hereafter "Lt Gen J Conway Holds Defense News Briefing."
282. I MEF Sitrep 271800ZMar03 to 281759ZMar03 (Copy in Reynolds Working Papers, MCHC, Quantico, VA).
283. I MEF Sitrep 281800ZMar03 to 291759ZMar03 (Copy in Reynolds Working Papers, MCHC, Quantico, VA).
284. "Lt Gen J Conway Holds Defense News Briefing."
285. I MEF Sitrep 271800ZMar03 to 281759ZMar03 (Copy in Reynolds Working Papers, MCHC, Quantico, VA). See also 1stMarDiv ComdC, Jan-Jun03 (GRC, Quantico, VA), sec 2, chap 6.
286. 2d MEB ComdC, Jan-Jun03 (GRC, Quantico, VA).
287. Mattis intvw no. 1.
288. Col Charles J. Quilter II, e-mail to author, 19Feb04 (Copy in Reynolds Working Papers, MCHC, Quantico, VA). See also I MEF Sitrep 301800ZMar03 to 311759Z Mar03 (Copy in Reynolds Working Papers, MCHC, Quantico, VA), containing a brief account of the Amos-Mattis meeting: "True to form, 3rd MAW pledged its utmost support to the Div's planned scheme of maneuver against the Baghdad Div, to include addressing critical resupply needs."
289. Col Charles J. Quilter II, e-mail to author, 21Feb04 (Copy in Reynolds Working Papers, MCHC, Quantico, VA); Stu Saffer, "Col Charlie Quilter, USMC, The Oldest Marine in Iraq," *Laguna Life and People* (Laguna Beach, CA: Mar04), p. 18.
290. The concept of operations is clearly outlined in I MEF Sitrep 01800ZMar03 to 311759ZMar03 (Copy in Reynolds Working Papers, MCHC, Quantico, VA). See also 1stMarDiv ComdC, Jan-Jun03 (MCHC, Quantico, VA), sec 2, chap 5.
291. Mattis intvw no. 1.
292. Ibid.; Toolan intvw; Conway intvw. It was reported in the press at the time. See, for example, Tony Perry, "Marine Commander Relieved of Duties," *Los Angeles Times*, 5Apr03, p. A-15. A later article offered a comprehensive picture of the affair: C. Cooper, "How a Marine Lost His Command." The History and Museums Division conducted an interview with Col Dowdy in December 2003.

Chapter 7

293. Fontenot, et al., *On Point*, pp. 49-50, provides an excellent discussion of this topic, which is the basis for this paragraph.
294. See Blackman intvw; 1stMarDiv ComdC, Jan-Jun03 (GRC, Quantico, VA), sec 2, chap 3.
295. McKiernan intvw no. 1. For further discussion, see for example, 1stMarDiv ComdC, Jan-Jun03 (GRC,

Quantico, VA), chap 3, p. 13; and Atkinson, *In the Company of Soldiers*, p. 26, quoting Army Brig. General Benjamin C. Freakley, the assistant division commander for operations of the 101st Airborne Division; West and Smith, *March Up*, p. 186, also discuss urban tactics.

296. See Reynolds, Journal, entries for 3-4Apr03; Col Thomas C. Latsko intvw, 22Apr03 (MCHC, Quantico, VA).

297. Paul Martin, "British Tactics in Basra Praised," *Washington Times*, 3Apr03, p. 1; Peter Baker, "Tactics Turn Unconventional; Commanders Draw Lessons of Belfast in Countering Attacks," *The Washington Post*, 20Mar03, p. A-23.

298. 1stMarDiv ComdC, Jan-Jun03 (GRC, Quantico, VA), sec 2, chap 3, pp. 19-20. For information on Urban Warrior, see Randolph Gangle intvw, 18Oct02 (MCHC, Quantico, VA), hereafter Gangle intvw.

299. Gangle intvw.

300. Mattis intvw no. 1.

301. 1stMarDiv ComdC, Jan-Jun03 (GRC, Quantico, VA), sec 2, chap 6; West and Smith, *March Up*, p. 207.

302. McKiernan intvws no. 1 and no. 4; Capt Matthew H. Bazarian intvw, 12Apr03 (MCHC, Quantico, VA), containing the comment that during Internal Look, General McKiernan made it clear he did not want to have two commands splitting Baghdad.

303. I MEF ComdC, Jan-Jun03 (GRC, Quantico, VA), sec 2, future operations command chronology and future operations folder, "CG Guidance/discussion with OPT 25 Mar 03"; Buhl intvw.

304. McKiernan intvw no. 4.

305. McKiernan intvw no. 1. For a slightly different view, see Acosta, Journal, entry for 20May03, reporting a briefing by the I MEF G-3 and his statement that "about a week before the war began, COMCFLCC gave I MEF the mission to secure eastern Baghdad." This is probably a slightly garbled reference to the branch plan.

306. I MEF ComdC, Jan-Jun03 (GRC, Quantico, VA) sec 2, future operations; Fontenot, et al., *On Point*, p. 331. Fontenot makes the point that flexibility was intentionally built into the plan.

307. "CFLCC Update for 3 Apr 03" (Disk 65, OIF-I Document Collection, GRC, Quantico, VA); CFLCC Briefings (Disks 55 and 56, OIF-I Document Collection, GRC, Quantico, VA), entries for 3-4Apr03.

308. Amos intvw.

309. 1stMarDiv ComdC, Jan-Jun03 (GRC, Quantico, VA), sec 2, chap 6.

310. The "digital divide" is explored in some detail in MCCDC, *OIF Summary Report*.

311. Milburn intvw no. 2.

312. For an excellent discussion of the BUA, see Maj Robert K. Casey intvw, 27Apr03 (MCHC, Quantico, VA). There is an equally good description of how the CFLCC staff worked in R. T. Johnson intvw. The author attended numerous BUAs in March and April 2003 and heard Generals McKiernan and Blackman talk about how they wanted them to work. See also Blackman intvw.

313. Judge intvws no. 1 and 2.

314. There are already numerous books and articles that address the subject of embedding. An excellent example is Atkinson, *In the Company of Soldiers*. On the Marine Corps side, there is Capt Dan McSweeney, "Clowns to the Left of Me," *U.S. Naval Institute Proceedings*, Nov03, pp. 46-48

315. I MEF Sitrep 051800ZApr03 to 061759ZApr03 (Copy in Reynolds Working Papers, MCHC, Quantico, VA).

316. West and Smith, *March Up*, p. 207. This was certainly consistent with General McKiernan's command style, which was to tell commanders what he wanted them to do and then leave the rest up to them.

317. McKiernan intvw no. 4; Fontenot, et al., *On Point*, chapter 6 offers a detailed discussion of the Army's movements.

318. P. J. Kennedy intvw.

319. I MEF Sitrep 051800ZApr03 to 061759ZApr03 (Copy in Reynolds Working Papers, MCHC, Quantico, VA).

320. I MEF Sitrep 061800ZApr03 to 071759ZApr03 (Copy in Reynolds Working Papers, MCHC, Quantico, VA).

321. I MEF Sitrep 071800ZApr03 to 081759ZApr03 (Copy in Reynolds Working Papers, MCHC, Quantico, VA).

322. 1stMarDiv ComdC, Jan-Jun03 (GRC, Quantico, VA), sec 2, chap 6, p. 41.

323. Ibid.

324. McKeldin, Journal, entry for 6Apr03; Judge intvw no. 1; Hummer intvw.

325. 1stMarDiv ComdC, Jan-Jun03 (GRC, Quantico, VA), sec 2, chap 6, p. 49; I MEF Sitrep 061800ZApr03 to 071759ZApr03 (Copy in Reynolds Working Papers, MCHC, Quantico, VA).

326. I MEF Sitrep 071800ZApr03 to 081759ZApr03 (Copy in Reynolds Working Papers, MCHC, Quantico, VA).

327. 1stMarDiv ComdC, Jan-Jun03 (GRC, Quantico, VA), sec 2, chap 6, p. 51.

328. McKeldin, Journal, entry for 9Apr03.

329. Peter Baker, "Marines' Orders: Ready, Set,

Switch," *The Washington Post*, 10Apr03, p. A-35, hereafter Baker, "Marines' Orders." Baker's report tracks closely with the I MEF sitrep for the same time period.
330. McKeldin, Journal, entry for 9Apr03.
331. 1stMarDiv ComdC, Jan-Jun03 (GRC, Quantico, VA), sec 2, chap 6, p. 50.
332. Capt Brian B. Smalley intvw, 3May03 (MCHC, Quantico, VA).
333. Quoted in Mark Mazzetti, "Lt General James Conway, Commander of the 1st Marine Expeditionary Force," *U.S. News and World Report* web exclusive, 9Apr03, p. 1.
334. Baker, "Marines' Orders."
335. Ibid.; Conway intvw; West and Smith, *March Up*, pp. 226-227, 233-234.
336. I MEF Sitrep 081800ZApr03 to 091759ZApr03 (Copy in Reynolds Working Papers, MCHC, Quantico, VA); Hummer intvw.
337. 1stMarDiv ComdC, Jan-Jun03 (GRC, Quantico, VA), sec 2, chap 6, p. 53; McKeldin, Journal, entry for 9Apr03; Fontenot, et al., *On Point*, pp. 337-338, quotes a good firsthand account.
338. Reynolds journal, entry for 9Apr03.
339. Dunford intvw; 1stMarDiv ComdC, Jan-Jun03 (GRC, Quantico, VA), sec 2, chap 6.
340. McKiernan intvw no. 4. See also G. W. Smith intvw no. 1, especially the comment that the Coalition spent a lot of time and effort working out how it would break things but not a lot on how it would put things back together.
341. Mazetti, "Lt General James Conway." See also Atkinson, *In the Company of Soldiers*, p. 287.
342. William Branigin and Anthony Shadid, "Authority Melts in Baghdad," *The Washington Post*, 9Apr03, p. A-1.
343. McKeldin, Journal, entry for 10Apr03; I MEF Sitrep 091800ZApr03 to 101759ZApr03 (Copy in Reynolds Working Papers, MCHC, Quantico, VA).
344. 1stMarDiv ComdC, Jan-Jun03 (GRC, Quantico, VA), sec 2, chap 6, p. 65.
345. I MEF Sitrep 091800ZApr03 to 101759ZApr03 (Copy in Reynolds Working Papers, MCHC, Quantico, VA).
346. 1stMarDiv ComdC, Jan-Jun03 (GRC, Quantico, VA), sec 2, chap 6, p. 65.
347. See, for example, Jehl, "Inspector's Report Says Hussein Expected Guerrilla War." This does not mean a guerrilla war was carefully planned.
348. 2d MEB ComdC, Jan-Jun03 (GRC, Quantico, VA), sec 2; Text of Cmdr, MarForPac, "I MEF (Rein) Unit Award Recommendation for Presidential Unit Citation," covering the period 21Mar-24Apr03.

Chapter 8

349. Mazetti, "Lt General James Conway." See also Conway intvw for a report of his discussions with CFLCC on the north. For an excellent discussion of the strategic dimension, see McKiernan intvw no. 1.
350. McKeldin, Journal, entry for 14Apr03.
351. McKeldin, Journal, entry for 10Apr03; I MEF Sitreps 081800ZApr03 to 091759ZApr03 and 091800ZApr03 to 101759ZApr03 (Copies in Reynolds Working Papers, MCHC, Quantico, VA).
352. I MEF Sitrep 071800ZApr03 to 081759ZApr03 (Copy in Reynolds Working Papers, MCHC, Quantico, VA).
353. West and Smith, *March Up*, p. 247, report that in September 2002, General Mattis ordered a contingency plan for a quick-moving task force to assist the Kurds if Turkey did not agree to open a northern front.
354. I MEF Sitrep 091800ZApr03 to 101759ZApr03 (Copy in Reynolds Working Papers, MCHC, Quantico, VA); 1stMarDiv ComdC, Jan-Jun03 (GRC, Quantico, VA), sec 2, chap 7. Except where indicated, the rest of this chapter is based on these two sources. The oral histories at MCHC offer additional detail, such as that of Capt Brian B. Smalley on 3 May 2003, who provides a vivid description of this operation from a company commander's point of view.
355. Amos intvw.
356. McKeldin, Journal, entry for 12Apr03; Conway intvw.
357. I MEF Sitrep 121800ZApr03 to 131759ZApr03 (Copy in Reynolds Working Papers, MCHC, Quantico, VA).
358. 1stMarDiv ComdC, Jan-Jun03 (GRC, Quantico, VA), sec 2, chap 7, p. 21.
359. McKeldin, Journal, entry for 12Apr03.
360. I MEF Sitreps 121800ZApr03 to 131759ZApr03, 141800ZApr03 to 151759ZApr03 (Copies in Reynolds Working Papers, MCHC, Quantico, VA).
361. I MEF Sitrep 181800ZApr03 to 191759ZApr03 (Copy in Reynolds Working Papers, MCHC, Quantico, VA).
362. 1stMarDiv ComdC, Jan-Jun03 (GRC, Quantico, VA), sec 2, chap 8, pp. 4, 6, discuss this phase of the operation.
363. Ibid.; and Reynolds, Journal, entry for 4May03. Col Christopher C. Conlin came to much the same conclusion in his article, "What do you do for an encore?" *Marine Corps Gazette*, Sep04, pp. 74, 80.
364. 1stMarDiv ComdC, Jan-Jun03 (GRC, Quantico, VA), sec 2, chap 8, p. 6.
365. Taken from the I MEF Casualty Report as of

14May03, reported in Acosta, Journal, entry for 20May03. These statistics include the casualties from the British division (20 KIA, 36 WIA). For the other MSCs, the breakdown was 1stMarDiv (22 KIA, 188 WIA); 1st FSSG (2 KIA, 2 WIA); 3d MAW (11 KIA, 8 WIA); Task Force Tarawa (19 KIA, 58 WIA).

Chapter 9

366. Karl Vick and Daniel Williams, "U.S. Troops Arrive in Kurd Area to Open Front," *The Washington Post*, 24Mar03, p. A-15.
367. MajGen Henry P. Osman intvw, 19Nov03 (MCHC, Quantico, VA).
368. Ibid. Murray and Scales, *Iraq War*, pp. 186, 195, is an excellent potted history of this campaign within a campaign. For information on Provide Comfort, as well as general background, see LtCol Ronald J. Brown, *Humanitarian Operations in Northern Iraq, 1991: With Marines in Operation Provide Comfort* (Washington, DC: History and Museums Division, 1995).
369. Jones intvw.
370. Col Robert L. Hayes III intvw, 19Nov03 (MCHC, Quantico, VA); Osman intvw; Zinni intvw.
371. Hayes intvw; LtCol James E. Bacchus intvw, 20Nov03 (MCHC, Quantico, VA), hereafter Bacchus intvw.
372. David Josar, "Marines, Army to Coordinate Humanitarian Aid," *European Stars and Stripes*, 30Mar03.
373. Bacchus intvw; Notes on meeting with Col Keith A. Lawless, 25Nov03 (Copy in Reynolds Working Papers, MCHC, Quantico, VA), hereafter Lawless notes. For the MCLC, reach back worked well, validating a 21st century model for staffing.
374. Lawless notes.
375. Gen H. P. Osman sitrep to Gen J. P. Abizaid, 7Apr03 (Copy in Reynolds Working Papers, MCHC, Quantico, VA).
376. Bacchus intvw.
377. Osman sitrep for 24Mar03.
378. Dan Williams and Philip P. Pan, "U.S. Plans to Create Military Command in Northern Iraq," *The Washington Post*, 25Mar03, p. 20.
379. Lawless notes.
380. Osman intvw.
381. Osman sitreps for 12, 13Apr03.
382. See Col John P. Holden intvw, 9Jun03 (Naval Historical Center, Washington, DC). At the time, Col Holden was serving as the deputy chief of staff for plans for the Sixth Fleet.
383. Col Andrew P. Frick intvw, 12Sep03 (MCHC, Quantico, VA).
384. Capt Arnaldo L. Colon, ed., "U.S. Marines in North Iraq, A Certain Force," n.d. (filed with 26th MEU ComdC, Jan-Jun03 (GRC, Quantico, VA).
385. Frick intvw. Colon, "A Certain Force," describes the command arrangement in some detail and is generally a very good source like a command chronology narrative. It tracks closely with my other basic source on 26th MEU, the narrative summary in the command chronology, 1Jan03-30Jun03, but includes more detail.
386. 1stLt Sunny-James M. Risler intvw, 15Oct03 (MCHC, Quantico, VA).
387. Sgt Bryan L. Gilstrap intvw, 16Oct03 (MCHC, Quantico, VA).
388. For a general reference with mentions of the incident, the allegations, and the general mood of the city, see David Rohde, "Deadly Unrest Leaves a Town Bitter at U.S.," *The New York Times*, 20Apr03, p. A-1.
389. Colon, "A Certain Force."
390. Frick intvw.
391. Osman sitrep for 22Apr03.
392. Jones intvw.

Chapter 10

393. House of Commons, *Lessons of Iraq*, v. 1, p. 45. As noted, this is a great source, full of direct quotations, without the "happy" feel of many British and American lessons-learned reports.
394. McKiernan intvw no. 1. See Huelfer intvw for evidence of early consideration of this option by CFLCC planners.
395. House of Commons, *Lessons of Iraq*, v. 1, p. 45.
396. R. T. Johnson intvw; Col Jeremy M. F. Robbins RM, intvw, 16Apr03 (MCHC, Quantico, VA); Murray and Scales, *Iraq War*, pp. 135-136. Murray and Scales place the relationship between the British and the Americans in the Gulf in a larger context, discussing developments over the preceding decade. A recent arrival on the bookshelves, John Keegan's *The Iraq War* (New York, NY: Knopf, 2004) is said to contain useful data on the British division. Excerpts from the book on I MEF, however, contain inaccuracies and add little to the literature.
397. "Major General R. V. Brims CBE," www.nato.int/sfor/comssfor/commndsw/t000121a.htm (Copy in Reynolds Working Papers, MCHC, Quantico, VA).
398. Reynolds, Journal, entry for 22Apr03.
399. GOC Directive 1, 3Feb03 (Copy in Reynolds

Working Papers, (MCHC, Quantico, VA). House of Commons, *Lessons of Iraq*, v. 1, p. 195, describes the British "manoeuverist" approach in much the same terms as FMFM 1: "long at the heart of British defense doctrine it is 'one in which shattering the enemy's cohesion and will to fight, rather than his materiel, is paramount.'"
400. See, for examples, G. W. Smith intvw no. 2 and Hummer intvw.
401. Cmdr 1(UK) Armd Div's Diary, entry for 7Mar03 (Copy in Reynolds Working Papers, MCHC, Quantico, VA), hereafter Commander's Diary.
402. Conway intvw; Ministry of Defence, *Operations in Iraq: First Reflections* (London, UK: 7Jul03), p. 8.
403. House of Commons, *Lessons of Iraq*, v. 1, p. 92.
404. Ibid., p. 93.
405. Ibid., p. 94.
406. MajGen Robin V. Brims intvw, 10May03 (MCHC, Quantico, VA). To be sure, there were one or two instances of friction between British and American officers.
407. For an excellent discussion of this topic from the British point of view, see House of Commons, *Lessons of Iraq*, v. 1, pp. 193, 195. This report points out that the distance from being a maneuverist to espousing effects-based planning is not far, they are certainly consistent.
408. See, for example, Capt Tracey A. Morris intvw, 30Mar03 (MCHC, Quantico, VA).
409. Latsko intvw; LtCol Edward C. Quinonez intvw, 12May03 (MCHC, Quantico, VA). See also Ministry of Defence, *Operations in Iraq: Lessons for the Future* (London, UK: 11Dec03), p. 34: "the implications of maintaining contact and congruence with US technological and doctrinal advances should continue to be assessed."
410. Reynolds, Journal, entry for 22Apr03.
411. Jones intvw.
412. Col Nicholas E. Reynolds, "Brief by Maj Chris Parker," 11May03 (Copy in Reynolds Working Papers, MCHC, Quantico, VA).
413. Ibid.; Walker intvw; Maj. Cary J. Schorsch, "UK ASE Chronology," n.d. (10May 03?) (Copy in Reynolds Working Papers, MCHC, Quantico, VA).
414. Commander's Diary, entry for 20Mar03. The implication was that the Iraqi missiles were fired in retaliation for the U.S. missile strikes against Baghdad the night before.
415. Ministry of Defense, *Lessons for the Future*, p. 12.
416. Commander's Diary, entry for 22Mar03.
417. Ministry of Defence, *Lessons for the Future*, p. 25.
418. Commander's Diary, entry for 23Mar03.
419. House of Commons, *Lessons of Iraq*, p. 97.
420. Ibid.
421. Latsko intvw.
422. Max Hastings, "The real story is how we won," *Sunday Telegraph* (London), 7Sep03, p. 4; Murray and Scales, *Iraq War*, pp. 144, 153, gives a good overview of the situation in Basrah.
423. Hastings, "The real story."
424. Ibid.; Commander's Diary, entry for 26Mar03
425. House of Commons, *Lessons of Iraq*, p. 99.
426. Commander's Diary, entry for 26Mar03.
427. LtCol Donald C. Wilson intvw, 22Apr03 (MCHC, Quantico, VA), described the working of British intelligence.
428. See also Latsko intvw.
429. House of Commons, *Lessons of Iraq*, p. 98.
430. Commander's Diary, entry for 4Apr03.
431. Murray and Scales, *Iraq War*, pp. 150-151; Commander's Diary, entry for 5Apr03. See K. L. Vantran, "CentCom Officials Announce Capture of 'Chemical Ali'," Armed Forces Information Service, 21Aug03.
432. Commander's Diary, entry for 6Apr03. See also I MEF sitrep 061800ZApr03 to 071759ZApr03 (Copy in Reynolds Working Papers, MCHC, Quantico, VA).
433. Brims intvw; Latsko intvw.
434. Murray and Scales, *Iraq War*, p. 152.
435. House of Commons, *Lessons of Iraq*, pp. 149-150.
436. Ibid., p. 62.
437. Ibid., p. 153.
438. Ibid.
439. I MEF sitrep 061800ZApr03 to 071759ZApr03 (Copy in Reynolds Working Papers, MCHC, Quantico, VA).
440. House of Commons, *Lessons of Iraq*, p. 156. This led into a somewhat philosophical discussion of whether the division's actions and effects desired by the division were really in sync. The point was that the division had won the war, but was anyone on the British side set up to win the peace?

Chapter 11

441. Peter Baker, "Top Officers Fear Wide Civil Unrest; Bloodshed among Iraqis Could Create Challenge for Invading Troops," *The Washington Post*, 19Mar03, p. A-18.
442. Benson, "Brief at Naval War College." Benson was the senior planner at CFCCC. See, for example, Bacevich, "Modern Major General," p. 129.
443. Moran intvw. See also G. W. Smith intvw no. 2; BGen Stephen Hawkins, USA (CG of CJTF-IV) intvw,

14Mar03 (U.S. Army Center of Military History, Washington, DC); Benson, "Brief at Naval War College."
444. LtCol Brian K. McCrary intvw, 6Nov03 (MCHC, Quantico, VA).
445. G. W. Smith intvw no. 2.
446. Peter Baker, "U.S. Forces Will Redeploy into 3 Zones," *The Washington Post*, 16Apr03, p. A-31.
447. Michael R. Gordon and John Kifner, "U.S. Generals Meet in Palace, Sealing Victory," *The New York Times*, 17Apr03.
448. I MEF sitrep 141800ZApr03 to 151759ZApr03 (Copy in Reynolds Working Papers, MCHC, Quantico, VA), announcing the upcoming meeting.
449. I MEF ComdC, Jan-Jun03 (GRC, Quantico, VA) "I MEF Sequel ("Post Hostility Operations" to I MEF OpOrd 1003V "Basic Order" and "FutOps Command Chronology," Jan-Jun03.
450. I MEF sitrep 111800ZApr03 to 121759ZApr03 (Copy in Reynolds Working Papers, MCHC, Quantico, VA).
451. Margaret Warner, "A Marine's View," Public Broadcasting System Online News Hour, 26Sep03. Good sources on the division during this period are 1stMarDiv ComdC, Jan-Jun03 (GRC, Quantico, VA), sec 2, chap 8; BGen John F. Kelly, "Tikrit, South to Babylon," *Marine Corps Gazette*, Feb-Apr04.
452. Max Boot, "Reconstructing Iraq," *Weekly Standard*, 15Sep03.
453. LtCol Patrick J. Malay intvw, 23Aug03 (MCHC, Quantico, VA).
454. Reynolds, Journal, entry for 4May03.
455. See, for example, 2dLt Glen J. Bayliff intvw, 4May03 (MCHC, Quantico, VA); LtCol Christopher C. Conlin intvw, 24Aug03 (MCHC, Quantico, VA); LtCol Andrew Pappas intvw, 20Aug03 (MCHC, Quantico, VA); Maj Joseph A. Cabell intvw, 28Aug03 (MCHC, Quantico, VA). LtCol Conlin, "mayor" of An Najaf, described his experiences in that city, while LtCol Pappas, head of a counter-insurgency task force, expressed considerable frustration at the lack of support from CPA in restoring the infrastructure. Criticism of ORHA and CPA was almost universal among Marines interviewed by field historians. See, for additional examples, Maj David P. Holahan intvw, 6Nov03 (MCHC, Quantico, VA); LtCol Robert O. Sinclair intvw, 7Nov03 (MCHC, Quantico, VA). Both LtCol Conlin and Maj Cabell commented on how useful the reservists' skills, and temperament, were.
456. For a balanced discussion of this policy, see Peter Slevin, "Wrong Turn at a Postwar Crossroads?" *The Washington Post*, 20Nov03, p. A-1. This was a Pentagon decision, opposed by many in the field, including General Franks. Franks, *American Soldier*, p. 441.
457. I MEF ComdC, Jan-Jun03 (GRC, Quantico, VA), sec 2, "FutOps Command Chronology."
458. Pamela Hess, "Raid in Iraq's 'Indian Country'," 5Aug03, United Press International. Journalist Max Boot visited the same unit shortly after Ms. Hess. See Boot, "Reconstructing Iraq." LtCol Pappas has described his experiences in a comprehensive interview. Pappas intvw.
459. Pappas intvw.
460. Maj Steven B. Manber intvw, 21Aug03 (MCHC, Quantico, VA).
461. Pamela Hess, "General: Iraq Chem-Bio Arms Intel Wrong," United Press International, 30May03.
462. BGen John F. Kelly, "Part III: Tikrit, South to Babylon," *Marine Corps Gazette*, Apr04, p. 46.
463. M. Warner, "Marine's View." See also Kelly, "Part III: Tikrit," p. 43, who notes: "[O]ur efforts with the local population assisted us in all but eliminating violence by midsummer." Max Boot shared this view. Boot, "Reconstructing Iraq." For the dissenting civilian view, see Neil MacFarquhar, "In Najaf, A Sudden Anti-U.S. Storm," *The New York Times,* 21Jul03, who notes: "The lack of Iraqis involved in the reconstruction at all levels, widespread unemployment, and woefully inadequate means of communicating [with U.S. forces] . . . have combined to fuel an ever-higher level of frustration and anger about the American presence."
464. 1st FSSG ComdC, Jul-Dec03 (GRC, Quantico, VA), sec. 2, "Current Ops Narrative Summary"; I MEF ComdC, Jul-Dec03 (GRC, Quantico, VA), sec. 2, "S-3 Operations and Training Narrative Summary"; Sgt Matthew Miller, "1st FSSG Goes Home," 1Oct03 (Story No. 20031015350, posted on MarineLink).
465. BGen Ronald S. Coleman intvw, 20Apr04 (MCHC, Quantico, VA).
466. Ibid.

Appendix A

Data Collection and Lessons Learned Process

Although history is not about lessons learned, even an overview of the first phase of the Iraq War would be incomplete without a few words about data collection. Not only did various teams collect a great deal of material that may be of use to future historians, but the process itself has an interesting development, not to mention some of the lessons learned themselves.

There was a time, especially after the combat phase, when it seemed that not enough tent space and computer terminals existed for all of the lessons-learned teams in theater. The senior lessons-learned team was from Joint Forces Command in Norfolk, tasked by the Pentagon to produce the official joint report. The Army had at least two lessons-learned groups in theater, including one whose officers produced the admirable preliminary Army history of Operation Iraqi Freedom, titled *On Point* (published by the Combat Studies Institute at Fort Leavenworth in 2004). The Marine Corps had the combat assessment team from the Marine Corps Combat Development Command at Quantico, Virginia. Then there were the various groups of field historians, embedded with "supported" commands. Once deployed, Marine historians worked closely with historians from other branches, especially their counterparts from the Army and the joint history staffs, as well as the Marine assessors and, to a lesser extent, other Services. Though there was a distinct pecking order among these various groups, with historians generally coming from organizations with relatively little bureaucratic clout, most got along well. Data was usually shared freely across the board. There has probably never been so much available, retrievable, and useful historical data.[1]

The recent history of the Marine assessors goes back to Desert Storm, when battle assessment teams deployed from the Marine Corps Combat Development Command to theater to conduct interviews and generally gather data to drive postwar analyses and complement the young Marine Corps Lessons Learned System, described as "a passive system," which relied on units to report their observations. Most of their interviews were anonymous, which limited their usefulness to historians. Their reports tended to go into established "channels," that is, they did not necessarily turn into front-burner action items. Today, more than 250 four-inch binders of Desert Storm material sit quietly on the shelves of the Gray Research Center at Quantico, and Marine Corps Lessons Learned System is largely unknown to many parts of the Marine Corps public. After 11 September, Lieutenant General Edward Hanlon, Jr., commanding Marine Corps Combat Development Command, ordered the creation of a combat assessment team that deployed to theater for Afghanistan to conduct data during, not after, combat operations and to turn it around quickly, in useful form, to the advocates, that is, representatives of the various communities of operators in the Marine Corps. Afghanistan was a useful opportunity to discover good and bad ways to learn lessons, and the result, less than a year later, was a sophisticated operation for the Iraq War led by Colonel Philip J. Exner, a dynamic thinker and operator out of the Combat Development Command's Studies and Analysis Division. He began by surveying the process:

> We . . . looked at past "lessons learned" efforts. Both authors and audiences were somewhat skeptical of the value of traditional approaches, which usually involved publication of a large tome or collection of documents. . . . One of the other services published an after action for a more recent operations that consisted of a 5,000-page main report with an 800-page executive summary and nearly 100,000 pages of appendices. Such monumental efforts often miss the very change agents who are essential to converting lessons into lessons learned because the action officers and decision-makers are often overwhelmed with information and chronically short of time.[2]

With the support of the Commandant of the Marine Corps, Exner worked to embed his assessors in the operating forces for the duration, in much the same way that journalists and historians were embedded, so that they could develop better access and understanding and collect better data. The data, in the form of interviews and surveys, went into a massive database in Quantico that was searchable and, to some extent, linked with data collected by historians working alongside the assessors. The assessment teams gave briefings to general officers, posted their findings on Web sites, and produced timely, reader-friendly reports for the advocates. For the more general Marine Corps public, Combat Development Command published the teams' findings in mercifully brief summary reports that initially were limited to official use but will no doubt find their ways into libraries and research centers before long.

This is not to say that the operators themselves neglected the after-action process. What was probably the first comprehensive "hot wash" took place in Bahrain at MarCent headquarters on 20 May 2003. The highlight was a blow-by-blow, chronological review of the operation by the I MEF operations officer, Colonel Larry K. Brown.[3] The next major evolution was sponsored by the Marine Corps Association at Quantico on 29 July 2003 and featured briefings by Generals James Conway, James Mattis, James Amos, and Richard Natonski.[4] General Conway's brief was a very good "executive summary" of the operation. Some of the specific objectives learned, covered on that day, were:

> The utility of the medical surgical units at the front; the use of SAPI, or small arms protective inserts, for the flak jackets; the positive impact of the embedded media; the concept of combat maintenance being performed with units on the fly; and the merit in organizing large, flexible combat battalions. . . . Challenges [that is, problem areas] included . . . [having two] Marine Corps supply systems (ATLASS I and ATLASS II), integration with special operations forces, casualty reporting, combat identification to prevent fratricide, and the need to sharpen Coalition intelligence sharing.[5]

A few weeks later, on 4 September 2003 there was another conference along the same lines, the Marine Corps Association and U.S. Naval Institute Forum 2003, which featured talks by General Mattis and retired Marine General Anthony Zinni, who was not afraid to strike out on his own and offer some pithy comments about how he saw the situation in Iraq. On 4 September, and then again in May 2004, Zinni criticized the Bush Administration's policy in language a drill instructor might have used; he said he remembered the official "garbage and lies" during the Vietnam era and asked if it was happening again. Zinni also offered a thoughtful analysis of what had gone wrong, the "10 mistakes" that "history is going to record." He attacked the reasons for going to war, arguing that the United States had successfully contained Saddam and that it needed first to resolve the conflict between Israel and Palestine. "I could not believe what I was hearing about the benefits of this strategic move. That the road to Jerusalem led through Baghdad, when just the opposite is true." He went on to make a number of points about Phase IV, citing the inadequate planning at both the Pentagon and CentCom levels, the inadequate number of troops for occupation duties, and the, to him amazing, decision to disband the Iraqi Army.[6]

Marines are likely to remember the observations about Phase III that emerged from these sessions. The first had to do with maneuver warfare and the Marine air-ground task force concept. There was general agreement that the Iraq War had revalidated Marine doctrine in at least two respects, speed and organization. The I Marine Expeditionary Force was organized and equipped for speed. It had moved much faster than the enemy; the enemy never had time to visualize the outlines of our "observation/orientation/decision/action" loop, let alone get inside it. The 3d Marine Aircraft Wing had remarkable new precision technologies (and new doctrines to go with them) to enable it to fight with unprecedented effectiveness. The wing and the 1st Force Service Support Group had not just supported division, which was itself organized into mobile, independent combat teams; they also had been maneuver elements in their own right, integrated into the overall scheme of maneuver. With the Marine Logistics Command's and the 1st Force Service Support Group's contributions, and the wing's willingness to switch from the deep battle to close air support to cargo missions on short notice, the division had been able to go the distance, to project Marine power on the ground far from saltwater. This was another way of saying that Operation Iraqi Freedom had revalidated the concept of the "MEF single battle." The I Marine Expeditionary Force had demonstrated, yet again, that the whole was greater than the sum of its parts, whether the issue was deep fires, rear area security, or keeping the supplies flowing to the front.[7]

Conclusions about Phase IV were more cautious. No one contradicted General Zinni and claimed there had been elaborate preparations for Phase IV. Nowhere in CentCom or Coalition Forces Land Component Command had there been a plan for Phase IV that was like the plan for Phase III, let alone all of the preparations that accompanied it, including the cross talk during its development, the many rehearsal of concept drills, and the exchange of liaison officers.

There were the arguments, like General Zinni's, for bringing many more American troops to theater for occupation service. He wanted them to be on hand before anyone crossed the line of departure, available to stabilize the country as soon as the fighting ended. A corollary advanced by some was that the Coalition could have moved more slowly from Kuwait to Baghdad in order to secure the objectives that had just been seized. Still, General Mattis repeated his assertion that he had had the right force mix on the ground in the summer of 2003: a battalion for each province, some aviation, and not much by way of mechanized assets. It was not necessarily how many troops there were on the ground, but what their skills were and what they were being told to do. That was why he had sent his mechanized Marines home in May. The mostly infantry Marines who stayed through the summer quickly proved their ability to shift and learn on the fly, and they did a more than creditable job as interim occupiers in the southern half of Iraq during the relatively brief period between the end of combat operations and their return home. Like General Mattis, at least one other senior I MEF officer stressed that one of the keys to success was getting the timing right, the longer the occupiers stayed, the greater the challenges would become. The implication was that the Marines succeeded in the short term but that any occupier would face problems in the medium and long term.[8]

Generals Earl Hailston, James Conway, James Mattis, and Anthony Zinni made one overarching point that will find favor with historians. It was that since the Iraqi military was comparatively weak, and since every contingency is unique in its own way, it is dangerous to over generalize from the Iraq War, to imagine that the next war will necessarily be like the last.

---

1. Two excellent sources about the overall lessons-learned process are James Jay Carafano, "After Iraq: Learning the War's Lessons," Heritage Foundation Backgrounder, No. 1664, 3Jul03, and Col Mark Cancian, "Learning the Lessons of War," 2004, unpublished article (Copy in Reynolds Working Papers, MCHC, Quantico, VA). On the Marines in particular, see LtGen Edward Hanlon, Jr., "Leaning Into the 21st Century," *Marine Corps Gazette*, Oct03, pp. 15, 17.

2. MCCDC, *OIF Summary Report*, p. 5. For further information, see MCCDC, "Memorandum for the Commandant of the Marine Corps/Battle Assessment Proposal," 19Jan91, and Officer-in-Charge, MarCent Assessment Team, "Letter of Instruction (Draft)," 3Dec01 (Copies in Reynolds Working Papers, MCHC, Quantico, VA).

3. Acosta, Journal, entry for 20May03.

4. Reynolds, Journal, entry for 29Jul03; Col John P. Glasgow, Jr., (Ret), "Editorial," *Marine Corps Gazette*, Sep03, p. 2.

5. Glasgow, "Editorial."

6. Thomas E. Ricks, "Ex-Envoy Criticizes Bush's Postwar Policy," *The Washington Post*, 5Sep03, p. A-13; Reynolds, Journal, entry for 5Sep03. For a more comprehensive look at lessons learned by Gen Zinni, see Anthony Zinni, "Ten Mistakes History Will Record about War in Iraq," *Defense Monitor*, v. XXXIII, No. 3 (May/Jun04), p. 1.

7. The focus here is on Marine lessons learned. The report of the Joint Forces Command's Joint Center for Lessons Learned highlighted achievements in the joint arena and, except for the issue of fratricide, was generally complimentary about the "jointness" of OIF. See, for example, Vernon Loeb, "Pentagon Credits Success in Iraq War to Joint Operations," *The Washington Post*, 3Oct03, p. A-15.

8. This is not necessarily an argument against medium- or long-term occupations. See, for example, Gunther intvw. With respect to troop strengths, some lessons-learned analysts disagreed with Gen Mattis' point of view and argued that especially for Phase IV the force had to be much heavier than it had been; the Army Chief of Staff, Gen Eric Shinseki, made the famous comment that it would take some 400,000 troops to occupy Iraq, and that it would have been better to sacrifice some speed in order to have deployed more force. A slower, heavier force might have gotten the job done better than the fast, light force that conquered Iraq in 21 days; the argument, which is generally inconsistent with current Marine thinking about how to fight the Marine air-ground task force, is that since the Iraqis were unable to put up much of a fight, it would have been better to proceed more methodically, securing and occupying terrain as the Coalition moved forward. This is one of the general implications of Fontenot, et al., *On Point*, and of the 3d Infantry Division's after-action report. See, for example, John L. Lumpkin and Dafna Linzer, "Army: Plan for Iraq Flawed," *Hartford Journal*, 28Nov03, p. A-1. Two thoughtful articles that explore the background to Phase IV in more depth are George Packer, "Letter from Baghdad: War after the War; What Washington Doesn't See in Iraq," *New Yorker*, 24Nov03, pp. 59, 85; Tom Donnelly and Gary Schmitt, "The Right Fight Now," *The Washington Post*, 26Oct03, p. B-1. Packer's article addresses the general topic of postwar reconstructions in the 21st century, and concludes that (a) it is lengthy process, (b) it is better when internationalized, and (c) the foundation of success is security. He then goes on to discuss the Pentagon's initial decision not to plan for a long-term occupation. Donnelly and Schmitt highlight the Marines' *Small Wars Manual*, which they say is as good a guide as any to postwar reconstruction.

Appendix B

Command List

U.S. Marine Forces, Central Command
March–November 2003*

Commanding General: LtGen Earl B. Hailston
Chief of Staff: Col Peter T. Miller
G-1: Col Richard B. Harris
G-2: Col William E. Rizzio
G-3: Col Timothy C. Wells
G-4: Col Philip N. Yff
G-5: Col Timothy L. Hunter
G-6: Col Kevin B. Jordan
SJA: LtCol Robert E. Pinder
Comptroller: Col Robert J. Herkenham

Combined/Joint Task Force-Consequence Management
Commanding General: BGen Cornell A. Wilson

Marine Logistics Command
Commanding General: BGen Michael R. Lehnert

Special Purpose MAGTF
Commanding General: BGen Ronald S. Coleman

I Marine Expeditionary Force (Reinforced)

Commanding General: LtGen James T. Conway
Deputy: MajGen Keith J. Stalder
Chief of Staff: Col John C. Coleman
G-1: Col William J. Hartig
G-2: Col Alan R. Baldwin
Col James R. Howcroft
G-3: Col Larry K. Brown
G-4: Col Matthew W. Blackledge
G-5: Col Christopher J. Gunther
Col Anthony L. Jackson
G-6: Col George J. Allen (to 15 June)
Col Marshall I. Considine (after 30 June)
SJA: Col William D. Durrett
Surgeon: Captain Joel A. Lees, USN
Chaplain: Captain John S. Gwudz, USN
I MEF Headquarters Group: Col John T. Cunnings
15th Marine Expeditionary Unit (SOC)
Commanding Officer: Col Thomas D. Waldhauser
24th Marine Expeditionary Unit (SOC)
Commanding Officer: Col Richard P. Mills (to 6 June)
Col Ronald J. Johnson (after 6 June)
11th Marine Expeditionary Unit, Command Element (-)

Commanding Officer: Col Anthony M. Haslam

Marine Ground Combat Element

1st Marine Division (Reinforced)
Commanding General: MajGen James N. Mattis
Assistant Division Commander: BGen John F. Kelly
Chief of Staff: Col Bennett W. Saylor
Col Joseph F. Dunford

1st Marine Regiment (-)(Reinforced) (Regimental Combat Team 1)
Commanding Officer: Col Joseph D. Dowdy (to 4 April)
Col John A. Toolan (after 4 April)

5th Marine Regiment (-) (Reinforced) (Regimental Combat Team 5)
Commanding Officer: Col Joseph F. Dunford (to 23 May)
Col R. Stewart Navarre (after 23 May)

7th Marine Regiment (-) (Reinforced) (Regimental Combat Team 7)
Commanding Officer: Col Steven A. Hummer

11th Marine Regiment (-)(Reinforced)
Commanding Officer: Col Michael P. Marletto

2d Marine Expeditionary Brigade (Task Force Tarawa)
Commanding General: BGen Richard F. Natonski
Chief of Staff: Col James W. Smoot

2d Marine Regiment (-) (Reinforced) (Regimental Combat Team 2)
Commanding Officer: Col Ronald L. Bailey

Marine Aviation Combat Element

3d Marine Aircraft Wing
Commanding General: MajGen James F. Amos
Assistant Wing Commander: BGen Terry G. Robling
Chief of Staff: Col Gerald A. Yingling, Jr.

Marine Aircraft Group 11 (-) (Reinforced)
Commanding Officer: Col Randolph D. Alles

Marine Aircraft Group 13 (-) (Reinforced)
Commanding Officer: Col Mark R. Saverese

Marine Aircraft Group 16 (-) (Reinforced)
Commanding Officer: Col Stuart L. Knoll

Marine Aircraft Group 29 (-) (Reinforced)
Commanding Officer: Col Robert E. Milstead, Jr.

Marine Wing Support Group 37 (-) (Reinforced)
Commanding Officer: Col Michael C. Anderson
Marine Air Control Group 38 (-) (Reinforced)
Commanding Officer: Col Ronnell R. McFarland

Marine Aircraft Group 39 (-)(Reinforced
Commanding Officer: Col Richard W. Spender
Col Kenneth P. Gardiner

Marine Combat Service Support Element

1st Force Service Support Group
Commanding General: BGen Edward G. Usher III
BGen Richard S. Kramlich

Deputy Commander: Col John L. Sweeney, Jr.
Chief of Staff: Col Darrell L. Moore

Combat Service Support Group 16 (Headquarters Elements)
Commanding Officer: LtCol Michael J. Taylor

Combat Service Support Group 11 (Brigade Service Support Group G 1)
Commanding Officer: Col John J. Pomfret
Col Charles L. Hudson

Combat Service Support Battalion 13 (4th Landing Support Battalion)
Commanding Officer: LtCol Michael D. Malone

Combat Service Support Group 14 (4th Supply Battalion)
Commanding Officer: Col John T. Larson

Combat Service Support Group 15 (1st Supply Battalion)
Commanding Officer: Col Bruce E. Bissett

Transportation Support Group
Commanding Officer: Col David G. Reist

I Marine Expeditionary Force Engineer Group
Commanding Officer: RAdm (UH) Charles R. Kubic, USN

United Kingdom Forces

1 Armoured Division (UK) (-)(Reinforced)
Commanding General: MajGen Robin V. Brims
Chief of Staff: Col Patrick Marriott

7 Armored Brigade (British Army)
Commanding Officer: Brig Graham Binns

16 Air Assault Brigade (British Army)
Commanding Officer: Brig Jacko Page

3 Commando Brigade, Royal Marines (-)
Commanding Officer: Brig Jim Dutton

---

*Includes billets in units which served in theater for part but not all of the period covered. Basic sources are MarAdmin 507/03, various versions, Oct-Dec03, with "Modifications to the I MEF Presidential Unit Citation Unit Listing," and unit command chronologies.

Appendix C:

Unit List

U.S. Marines In Operation Iraqi Freedom
March-November 2003*

U.S. Marine Forces, Central Command [USMarCent]

Command Element
Combined Joint Task Force-Consequence Management [CJTF-CM]
Marine Corps Logistics Command [MarLogCom]
Special Purpose Marine Air-Ground Task Force [SPMAGTF]

I Marine Expeditionary Force (Reinforced) [I MEF]

Command Element

15th Marine Expeditionary Unit (Special Operations Capable) [15th MEU (SOC)]

Battalion Landing Team 2d Battalion, 1st Marines [BLT 2/1]
Marine Medium Helicopter Squadron 161 (Reinforced) [HMM-161]
Marine Expeditionary Unit Service Support Group 15 [MSSG-15]

24th Marine Expeditionary Unit (Special Operations Capable) [24th MEU (SOC)]

Battalion Landing Team 2d Battalion, 2d Marines [BLT 2/2]
Marine Medium Helicopter Squadron 263 (Reinforced) [HMM-263]
Marine Expeditionary Unit Service Support Group 24 [MSSG-24]

Task Force Yankee [TF Yankee]

11th Marine Expeditionary Unit, Command Element (-) [11th MEU, CmdEle]
2d Battalion, 6th Marines (-) (Reinforced) [2d Bn, 6th Mar]
Sensitive Site Team Number 3, U.S. Army [SenSiteTm #3, USA]
75th Exploitation Task Force, U.S. Army [75th ExpTF, USA]
Company C, 478th Engineer Battalion, U.S. Army [Co C, 478th EngrBn, USA]

I Marine Expeditionary Force Headquarters Group [I MEF HqGru]

6th Communications Battalion (-) [6th CommBn]
9th Communications Battalion (-) [9th CommBn]
1st Radio Battalion (-) (Reinforced) [1st RadBn]
1st Intelligence Battalion (-) (Reinforced) [1st IntelBn]
1st Force Reconnaissance Company (-) (Reinforced) [1st ForReconCo]
I Marine Expeditionary Force Liaison Element [I MEF LsnEle]
3d Air Naval Gunfire Liaison Company [3d ANGLICO]
4th Air Naval Gunfire Liaison Company [4th ANGLICO]
3d Civil Affairs Group [3d CAG]
4th Civil Affairs Group [4th CAG]

Marine Ground Combat Element

1st Marine Division (Reinforced) [1st MarDiv]

Headquarters Battalion [HqBn]

1st Marines (-) (Reinforced)/Regimental Combat Team 1 [1st Mar/RCT-1]

Headquarters Company [HqCo]
3d Battalion, 1st Marines [3d Bn, 1st Mar]
1st Battalion, 4th Marines [1st Bn, 1st Mar]
2d Battalion, 23d Marines [2d Bn, 23d Mar]
2d Light Armored Reconnaissance Battalion (-) [2d LARBn]

5th Marines (-) (Reinforced)/Regimental Combat Team 5 [5th Mar/RCT-5]

Headquarters Company [HqCo]
1st Battalion, 5th Marines [1st Bn, 5th Mar]
2d Battalion, 5th Marines [2d Bn, 5th Mar]
3d Battalion, 5th Marines [3d Bn, 5th Mar]
2d Tank Battalion (-) (Reinforced) [2d TkBn]
1st Light Armored Reconnaissance Battalion (-) [1st LARBn]
Company C, 4th Combat Engineer Battalion [Co C, 4th CbtEngrBn]

7th Marines (-) (Reinforced)/Regimental Combat Team 7 [7th Mar/RCT-7]

Headquarters Company [HqCo]
1st Battalion, 7th Marines [1st Bn, 7th Mar]
3d Battalion, 7th Marines [3d Bn, 7th Mar]
3d Battalion, 4th Marines [3d Bn, 4th Mar]
1st Tank Battalion (-) (Reinforced) [1st TkBn]
3d Light Armored Reconnaissance Battalion, (-) (Reinforced) [3d LARBn]

11th Marines (-) (Reinforced) [11th Mar]

Headquarters Battery (-) [HqBtry]
Detachment, Headquarters Battery, 10th Marines [Det, HqBtry, 10th Mar]
1st Battalion, 11th Marines (-) [1st Bn, 11th Mar]
2d Battalion, 11th Marines (-) [2d Bn, 11th Mar]
3d Battalion, 11th Marines (-) [3d Bn, 11th Mar]
5th Battalion, 11th Marines (-) [5th Bn, 11th Mar]
1st Combat Engineer Battalion (-) (Reinforced) [1st CbtEngrBn]
2d Combat Engineer Battalion (-) (Reinforced) [2d CbtEngrBn]
1st Reconnaissance Battalion (-) (Reinforced) [1st ReconBn]
2d Assault Amphibian Battalion (-) (Reinforced) [2d AABn]
3d Assault Amphibian Battalion (-) (Reinforced) [3d AABn]
4th Assault Amphibian Battalion [4th AABn]
2d Radio Battalion [2d RadBn]
Military Police Company, 4th Marine Division [MPCo, 4th MarDiv]
Communications Company, 4th Marine Division [CommCo, 4th MarDiv]
Battery I, 3d Battalion, 10th Marines [Btry I, 3d Bn, 10th Mar]
Battery R, 5th Battalion, 10th Marines [Btry R, 5th Bn, 10th Mar]

2d Marine Expeditionary Brigade (Task Force Tarawa) [2d MEB TF Tarawa]

Command Element
Detachment, II Marine Expeditionary Force Headquarters Group [Det, II MEF HqGru]
II Marine Expeditionary Force Liaison Element [II MEF LsnEle]
2d Battalion, 6th Marines (Originally with Task Force Yankee) [2d Bn, 6th Mar]
Company C, 4th Reconnaissance Battalion [Co C, 4th ReconBn]
2d Force Reconnaissance Company [2d ForReconCo]
2d Intelligence Battalion (-) [2d IntelBn]

2d Marines (-) (Reinforced)/Regimental Combat Team 2 [2d Mar/RCT-2]

Headquarters Company [HqCo]
1st Battalion, 2d Marines [1st Bn, 2d Mar]
3d Battalion, 2d Marines [3d Bn, 2d Mar]
2d Battalion, 8th Marines [2d Bn, 8th Mar]
1st Battalion, 10th Marines [1st Bn, 10th Mar]
Battery F, 2d Battalion, 10th Marines [Btry F, 2d Bn, 10th Mar]
Company A, 2d Combat Engineer Battalion [Co A, 2d CbtEngrBn]
Company A, 8th Tank Battalion [Co A, 8th TkBn]
Company C, 2d Light Armored Reconnaissance Battalion [Co C, 2d LARBn]
Company A, 2d Assault Amphibious Battalion [Co A, 2d AABn]
Company A, 2d Reconnaissance Battalion (Reinforced) [Co A, 2d ReconBn]

Marine Aviation Combat Element

3d Marine Aircraft Wing [3d MAW]

Marine Wing Headquarters Squadron 3 [MWHS-3]
Detachment, Marine Wing Headquarters Squadron 2 [Det, MWHS-2]

Marine Aircraft Group 11 (-) (Reinforced) [MAG-11]

Marine Aviation Logistics Squadron 11 (-) (Reinforced) [MALS-11]
Marine Aviation Logistics Squadron 14 (-) [MALS-14]
Detachment, Marine Aviation Logistics Squadron 31 [Det, MALS-31]
Marine Aerial Refueler Transport Squadron 352 (-) (Reinforced) [VMGR-352]
Detachment, Marine Aerial Refueler Transport Squadron 234 [Det, VMGR-234]
Detachment, Marine Aerial Refueler Transport Squadron 452 [Det, VMGR-452]
Marine Fighter Attack Squadron 232 [VMFA-232]
Marine Fighter Attack Squadron 251 [VMFA-251]
Marine All Weather Fighter Attack Squadron 121 [VMFA(AW)-121]
Marine All Weather Fighter Attack Squadron 225 [VMFA(AW)-225]
Marine All Weather Fighter Attack Squadron 533 [VMFA(AW)-533]
Marine Tactical Electronic Warfare Squadron 1 [VMAQ-1]
Marine Tactical Electronic Warfare Squadron 2 [VMAQ-2]

Marine Aircraft Group 13 (-) (Reinforced) [MAG-13]

Marine Aviation Logistics Squadron 13 (-) [MALS-13]
Marine Attack Squadron 211 (-) [VMA-211]
Marine Attack Squadron 214 [VMA-214]
Marine Attack Squadron 223 (-) [VMA-223]

Marine Attack Squadron 311 [VMA-311]
Marine Attack Squadron 542 [VMA-542]

Marine Aircraft Group 16 (-) (Reinforced) [MAG-16]

Marine Aviation Logistics Squadron 16 (-) [MALS-16]
Detachment, Marine Aviation Logistics Squadron 26 [Det, MALS-26]
Marine Medium Helicopter Squadron 163 [HMM-163]
Marine Heavy Helicopter Squadron 462 [HMH-462]
Marine Heavy Helicopter Squadron 465 [HMH-465]

Marine Aircraft Group 29 (-) (Reinforced) [MAG-29]

Marine Aviation Logistics Squadron 29 (-) [MALS-29]
Marine Medium Helicopter Squadron 162 [HMM-162]
Marine Medium Helicopter Squadron 365 (-) [HMM-365]
Marine Heavy Helicopter Squadron 464 [HMH-464]
Marine Light Attack Helicopter Squadron 269 [HMLA-269]

Marine Wing Support Group 37 (-) (Reinforced) [MWSG-37]

Marine Wing Support Squadron 271 [MWSS-271]
Marine Wing Support Squadron 272 [MWSS-272]
Marine Wing Support Squadron 371 [MWSS-371]
Marine Wing Support Squadron 372 [MWSS-372]
Marine Wing Support Squadron 373 [MWSS-373]
Company C, 1st Battalion, 24th Marines [Co C, 1st Bn, 24th Mar]
Detachment, Military Police, 4th Marine Air Wing [Det, MP, 4th MAW]

Marine Air Control Group 38 (-) (Reinforced) [MACG-38]

Air Traffic Control Detachment B, Marine Tactical Air Control Squadron 2 [ATCDet B, MTACS-2]
Marine Air Control Squadron 1 (Reinforced) [MACS-1]
Detachment, Marine Air Control Squadron 2 [Det, MACS-2]
Marine Wing Communications Squadron 28 (-) [MWCS-28]
Marine Wing Communications Squadron 38 (Reinforced) [MWCS-38]
Detachment, Marine Tactical Air Control Squadron 28 [Det, MTACS-28]
Marine Tactical Air Control Squadron 38 (Reinforced) [MTACS-38]
Detachment, Marine Tactical Air Control Squadron 48 [Det, MTACS-48]
Marine Air Support Squadron 1 [MASS-1]
Marine Air Support Squadron 3 (Reinforced) [MASS-3]
Battery B, 2d Low Altitude Air Defense Battalion [Btry B, 2d LAADBn]
3d Low Altitude Air Defense Battalion [3d LAADBn]
Detachments, Marine Air Support Squadron 6 [CA, MA Dets, MASS-6]
Marine Unmanned Aerial Vehicle Squadron 1 [VMU-1]
Marine Unmanned Aerial Vehicle Squadron 2 [VMU-2]

Marine Aircraft Group 39 (-) (Reinforced) [MAG-39]

Marine Aviation Logistics Squadron 39 (-) [MALS-39]
Marine Light Attack Helicopter Squadron 169 [HMLA-169]
Marine Light Attack Helicopter Squadron 267 [HMLA-267]
Marine Medium Helicopter Squadron 268 [HMM-268]

Marine Medium Helicopter Squadron 364 [HMM-364]
Marine Light Attack Helicopter Squadron 369 [HMLA-369]
Atlantic Ordnance, Command Expeditionary Force [LantOrd, CmdExpedFor]
Detachment, Headquarters and Headquarters Squadron, Marine Corps Air Station, Miramar [Det, HHS, MCAS Miramar]

Marine Combat Service Support Element

1st Force Service Support Group [1st FSSG]**

Detachment, Headquarters and Service Battalion [Det, H&SBn]

Combat Service Support Group 11 (Brigade Service Support Group 1) [CSSG 11]

Headquarters [Hq]
Combat Service Support Battalion 10 (Combat Service Support Group 1) [CSSB 10]
Combat Service Support Company 111 [CSSC 111]
Combat Service Support Company 115 [CSSC 115]
Combat Service Support Company 117 [CSSC 117]

Combat Service Support Group 13 (4th Landing Support Battalion) [CSSG 13]

Headquarters, 4th Landing Support Battalion [Hq, 4th LdgSptBn]
Combat Service Support Company 133 [CSSC 133]
Combat Service Support Company 134 [CSSC 134]
Combat Service Support Company 135 [CSSC 135]

Combat Service Support Group 14 (4th Supply Battalion) [CSSG 14]
4th Supply Battalion (-) [4th SupBn]

Combat Service Support Group 15 (1st Supply Battalion) [CSSG 15]
1st Supply Battalion (-) [1st SupBn]
Combat Service Support Battalion 12 (1st Maintenance Battalion) [CSSB 12]
Combat Service Support Battalion 18 (Headquarters and Service Battalion, 1st Force Service Support Group) [CSSB 18]
Combat Service Support Battalion 22 [CSSB 22]
Combat Service Support Company 151 [CSSC 151]

Transportation Support Group [TransSuptGru]
1st Transportation Support Battalion (-) [1st TSptBn]
6th Motor Transport Battalion [6th MTBn]

7th Engineer Support Battalion (-) (Reinforced) [7th EngrSptBn]
6th Engineer Support Battalion (-) (Reinforced) [6th EngrSptBn]
8th Engineer Support Battalion (-) (Reinforced) [8th EngrSptBn]
Communications Company, 4th Force Service Support Group [CommCo, 4th FSSG]
Mortuary Affairs Company, 4th Force Service Support Group [MortAffairsCo, 4th FSSG]
Company A, Military Police, 4th Force Service Support Group [Co A, MP, 4th FSSG]
Company B, Military Police, 4th Force Service Support Group [Co B, MP, 4th FSSG]
Company C, Military Police, 4th Force Service Support Group [Co C, MP, 4th FSSG]
Combat Service Support Battalion 16 (Combat Service Support Detachment 16) [CSSB 16]
Combat Service Support Battalion 19 (Marine Expeditionary Unit Service Support Group-11) [CSSB 19]

1st Dental Battalion (-) [1st DentBn]
Fleet Hospital Three, U.S. Navy [FH 3, USN]
Fleet Hospital Fifteen, U.S. Navy [FH 15, USN]
Preventive Medicine Unit, Navy Environmental Health Center [PM-MMART-5]
Preventive Medicine Unit, Navy Environmental Health Center [PM-MMART-2]
Health Services Battalion (1st Medical Battalion) [Health ServBn/1st MedBn]
Company A [Co A]
Company B [Co B]
Company C [Co C]
Company E [Co E]
Company F [Co F]

I Marine Expeditionary Force Engineer Group [I MEFEngrGru]
Command Element:
30th Naval Construction Regiment [30th NCR]
Naval Mobile Construction Battalion 5 [NMCB 5]
Naval Mobile Construction Battalion 7 [NMCB 7]
Naval Mobile Construction Battalion 74 [NMCB 74]
Naval Mobile Construction Battalion 133 [NMCB 133]
Naval Mobile Construction Battalion 4 [NMCB 4]
Naval Construction Force Support Unit 2 (-) [NCFSU 2]
Air Detachment, Underwater Construction Team 2 [AirDet, UCT 2]

22d Naval Construction Regiment [22d NCR]

Air Detachment, Naval Mobile Construction Battalion 15 [AirDet, NMCB 15]
Air Detachment, Naval Mobile Construction Battalion 21 [AirDet, NMCB 21]
Air Detachment, Naval Mobile Construction Battalion 25 [AirDet, NMCB 25]
Detachment, Construction Battalion Maintenance Unit 303 [Det, CBMU 303]

United Kingdom (UK) Forces

1 Armoured Division (UK) (-) (Reinforced) [1 ArmdDiv (UK)]
7 Armoured Brigade (UK) [7 ArmdBde (UK)]
1st Battalion, The Black Watch [1st Bn, BlackWatch]
1st Battalion, The Royal Regiment of Fusiliers [1st Bn, RoyalFusiliers]
The Royal Scots Dragoon Guards [RoyalScotsDragoons]
2d Royal Tank Regiment [2d Royal TkRegt]
3d Regiment, Royal Horse Artillery [3d Regt, RoyalHorseArty]
32 Engineer Regiment [32 EngrRegt]
16 Air Assault Brigade (UK) [16 AirAsltBde (UK)]
1st Battalion, The Parachute Regiment [1st Bn, ParaRegt]
3d Battalion, The Parachute Regiment [3d Bn, ParaRegt]
1st Battalion, The Royal Irish Regiment [1st Bn, RoyalIrishRgt]
7th Regiment, Royal Horse Artillery (Parachute) [7th Regt, RoyalHorseArty (Para)]
3 Commando Brigade, Royal Marines (-) [3 CdoBde, RM]
40 Commando Group [40 Cdo, RM]
42 Commando Group [42 Cdo, RM]
29 Commando Regiment, Royal Artillery [29 Cdo, RoyalArty]

U.S. Army Reinforcing Units

Detachment, 9th Psychological Operations Battalion [Det, 9th PsyOpsBn]

354th Public Affairs Detachment [354th PADet]
Detachment, Headquarters and Headquarters Company, 468th Chemical Battalion [Det, HHCo, 468th ChemBn]
U.S. Army Space Support Team [USASpaceSptTm]
86th Signal Battalion [86th SigBn]
208th Signal Company [208th SigCo]
Company C, 40th Signal Battalion [Co C, 40th SigBn]
3d Battalion, 27th Field Artillery [3d Bn, 27th FldArty]
1st Field Artillery Detachment [1st FldArtyDet]
498th Medical Company [498th MedCo]
Headquarters and Headquarters Battery, 108th Air Defense Artillery Brigade [HHBtry, 108th AirDefArtyBde]
2d Battalion, 43rd Air Defense Artillery [2d Bn, 43d AirDefArty]
3d Battalion, 124th Infantry [3d Bn, 124th Inf]
555th Maintenance Company [555th MaintCo]
Detachment, Headquarters and Headquarters Company, 378th Support Battalion [Det, HHCo, 378th SptBn]
777th Maintenance Company [777th MaintCo]
727th Transportation Company [727th TransCo]
319th Transportation Company [319th TransCo]
319th Transportation Detachment [319th TransDet]
299th Engineer Company [299th EngrCo]
459th Engineer Company [459th EngrCo]
Detachment, Headquarters and Headquarters Co., 716th Military Police Battalion [Det, HHCo, 716th MPBn]
Headquarters and Headquarters Company, 265th Engineer Group [HHCo, 265th EngrGru]
130th Engineer Battalion [130th EngrBn]
478th Engineer Battalion [478th EngrBn]
Headquarters and Headquarters Company, 358th Civil Affairs Brigade [HHCo, 358th CABde]
304th Civil Affairs Brigade [304th CABde]
402d Civil Affairs Brigade [402d CABde]
432d Civil Affairs Battalion [432d CABn]
367th Mobile Public Affairs Detachment [367th MOPADet]
305th Tactical Psychological Operations Company [305th TacPsyOpsCo]
307th Tactical Psychological Operations Company [307th TacPsyOpsCo]
312th Tactical Psychological Operations Company [312th TacPsyOpsCo]
1092d Engineer Battalion [1092d EngrBn]

Marine Follow-on Forces

3d Battalion, 23d Marines [3d Bn, 23d Mar]
4th Combat Engineer Battalion (-) [4th CbtEngrBn]
4th Light Armored Reconnaissance Battalion (-) [4th LARBn]
2d Battalion 25th Marines [2d Bn, 25th Mar]
Truck Company, 4th Marine Division [TkCo, 4th MarDiv]

Marine Forces with Fifth and Sixth Fleets

26th Marine Expeditionary Unit (Special Operations Capable) [26th MEU (SOC)]
  Battalion Landing Team 1st Battalion, 8th Marines [BLT 1/8]
  Marine Medium Helicopter Squadron 264 [HMM-264]
  Marine Expeditionary Unit Service Support Group 26 [MSSG 26]

Marine Fighter Attack Squadron 115 (USS *Harry S. Truman* CVN 75) [VMFA-115]
Marine Fighter Attack Squadron 312 (USS *Enterprise* CVN 65) [VMFA-312]
Marine Fighter Attack Squadron 323 (USS *Constellation* CV 64) [VMFA-323]

*Unit list based on I Marine Expeditionary Force Presidential Unit Citation Award Recommendation, 7Aug03; MarAdmin 507/03, various versions, Oct-Dec03; "Modifications to the I MEF Presidential Unit Citation Unit Listing," with additions and/or corrections provided by Ms. Annette Amerman, Historian, Reference Branch; and Col Nicholas E. Reynolds' troop list of Oct04. Unit abbreviations are provided in brackets.

**1st Force Service Support Group reorganized shortly before deployment; previous unit designations are shown in parentheses after the unit's designation in Operation Iraqi Freedom. The 1st Force Service Support Group's headquarters elements were reorganized into Combat Service Support Group 16 [CSSG 16] in April 2003.

Appendix D

Selected Glossary of Terms and Abbreviations

AA–Assault Amphibian
AAA–Antiaircraft Artillery
AAOE–Arrival and Assembly Operations Echelon
AAV–Amphibious Assault Vehicle
ACE–Aviation Combat Element
ADC–Assistant Division Commander
ADOCS–Automated Deep Operations Coordination System
APOD-Air Port of Debarkation
APOE–Air Port of Embarkation
ASLT–Air Support Liaison Team
ASOC–Air Support Operations Center
ASP–Ammunition Supply Point
ATARS–Advanced Tactical Air Reconnaissance System
ATO–Air Tasking Order
BCL–Battlefield Coordination Line
BCT–Brigade Combat Team
BDA–Battle Damage Assessment
BFT–Blue Force Tracker
BSSG–Brigade Service Support Group
C2PC–Command and Control Personal Computer
CBR–Counter Battery Radar
CE–Command Element
CEB–Combat Engineering Battalion
CENTCOM–U.S. Central Command
CFACC–Coalition Forces Air Component Commander
CFLCC–Coalition Forces Land Component Commander
CG–Commanding General
CGS–Common Ground Station
CIP–Combat Identification Panel
Class II–Batteries
Class VIII–Medical Supplies
Class IX–Repair Parts
CMOC–Civil-Military Operations Center
CPAO–Consolidated Public Affairs Office
CP-Command Post
CPX–Command Post Exercise
CRAF–Civil Reserve Air Fleet
CSS–Combat Service Support
CSSB–Combat Service Support Battalion

CSSC–Combat Service Support Company
CONPLAN–Contingency Plan
CONUS–Continental United States
COP–Common Operational Picture
DA–Dispersal Area
DAC–Division Administration Center
DASC–Direct Air Support Center
DIA–Defense Intelligence Agency
DOC–Deployment Operations Center
DS–Direct Support
DSA–Division Support Area
EMCON–Emissions Control
EOD–Explosive Ordnance Disposal
EPW–Enemy Prisoner of War
FAC–Forward Air Controller
FAD–Field Artillery Detachment
FARP–Forward Arming and Refueling Point
FOB–Forward Operating Base
FOE–Follow on Echelon
FPOL–Forward Passage of Lines
FRAGO–Fragmented Order
FRSS–Forward Resuscitative Surgery System
FSCC–Fire Support Coordination Center
FSS–Fast Sealift Ships
FSSG–Force Service Support Group
GBS–Global Broadcasting System
GCE–Ground Combat Element
GOSP–Gas-Oil Separation Plant
HDR–Humanitarian Daily Ration
HET–Human Exploitation Team
HF–High Frequency
HHA–Hand Held Assay
HUMINT–Human Intelligence
IC–Intelligence Community
IMINT–Image Intelligence
IMO–Information Management Officer
IO–Information Officer
IPSA–Intermediate Pumping Stations
JDAM–Joint Direct Attack Munition
JMEM–Joint Munitions Effectiveness Manual
JSTARS–Joint Surveillance and Target Attack Radar System
KAF–Kuwaiti Armed Forces
KI–Killbox Interdiction
KLF–Kuwaiti Land Forces
KMOD–Kuwaiti Ministry of Defense

LAR–Light Armored Reconnaissance
LASER–Light Amplification through Stimulated Emission of Radiation
LAV–Light Armored Vehicle
LD–Line of Departure
LOC–Line of Communication
LSA–Life Support Area; Logistical Support Area
LTO–Logistics Tasking Order
LZ–Landing Zone
MACCS–Marine Air Command and Control Squadron
MAG–Marine Air Group
MAGTF–Marine Air-Ground Task Force
MANPAD–Man-Portable Air Defense
MARCORSYSCOM–Marine Corps Systems Command
MAW–Marine Aircraft Wing
MCIA–Marine Corps Intelligence Activity
MCRE–Marine Corps Readiness Evaluation
MCWL–Marine Corps Warfighting Laboratory
MDACT–Mobile Data Automated Communication Terminal
MEB–Marine Expeditionary Brigade
MEF–Marine Expeditionary Force
MEFEX–Marine Expeditionary Force Exercise
MEG–MEF (Marine Expeditionary Force) Engineer Group
MEWSS–Mobile Electronic Warfare Support System
MLC–Marine Logistics Command
MOD–Ministry of Defense (Kuwait)
MOI–Ministry of the Interior (Kuwait)
MOPP–Mission Oriented Protective Posture
MOS–Military Occupational Specialty
MOUT–Military Operations on Urban Terrain
MP–Military Policy
MPF–Maritime Prepositional Force
MPSRON–Maritime Prepositioning Ship Squadron
MRLS–Multiply Rocket Launcher System
MSC–Major Subordinate Command
MSTP–MAGTF Staff Training Program
MWSG–Marine Wing Support Squadron
MWSS–Marine Wing Support Squadron
NBC–Nuclear, Biological, and Chemical
NBCRS–Nuclear, Biological, and Chemical Reconnaissance System
OCD–Obstacle Clearing Detachment
OMC-K Office of Military Cooperation-Kuwait
OPCON–Operation Control
OPLAN–Operations Plan
OPP–Offload Preparation Party
OPT–Operational Planning Team

ORCON–Originator Controlled
OSW–Operation Southern Watch
PA–Public Affairs
PALT–Public Affairs Liaison Team
PIR–Priority Intelligence Requirement
PLI–Position Location Information
POL–Passage of Lines
POW–Prisoner of War
PRR–Personal Role Radio
QRF–Quick Reaction Force
RA–Regular Army
RCT–Regimental Combat Team
RFF–Requested for Forces
RG–Republican Guard
RGFC–Republican Guard Forces Command
RIP–Relief in Place
ROC–Rehearsal of Concept
ROZ–Restrical Operation Zone
RRP–Refueling and Replenishment Point
RSO&I–Reception, Staging, Onward Movement and Integration
RUC–Reporting Unit Code
SAPOE–Sea and Aerial Ports of Embarkation
SAM–Surface-to-Air-Missile
SASO–Security and Stabilization Operations
SIGINT–Signal Intelligence
SIPRNET–Secret Internet Protocol Routed Network
SLTLP–Survey, Liaison, and Reconnaissance Party
SMART-T–Secure Mobile Antijam Reliable Tactcal Terminal
SOP–Standing Operating Procedure
SRG–Special Republican Guard
SPINS–Special Instructions
SPOD–Sea Port of Debarkation
SPOE–Sea Port of Embarkation
SSE–Sensitive Site Exploitation
SSM–Surface-to-Surface Missile
TAA–Tactical Assembly Areas
TACON–Tactical Control
T/E–Table of Equipment
TEWT–Tactical Exercise Without Troops
TIO–Target Information Officer
TIP–Thermal Identification Officer
T/O–Table of Organization
TPC–Target Procesing Center
TPFDD–Time-Phased Force Deployment Data

Appendix E

Chronology of Events

2001

11 September	Al Qaeda terrorists attack the World Trade Center and the Pentagon.
25 November	Marines of Task Force 58 land in Afghanistan as part of operations to deprive Al Qaeda of its base in that country.

2002

January	Marine Forces, Pacific, orders I Marine Expeditionary Force (I MEF) to focus on preparing for contingencies in the U.S. CentCom theater; I MEF planners begin more than a year of work on plans to invade Iraq.
2 August	MajGen James N. Mattis becomes Commanding General, 1st Marine Division, and puts the division on a virtual war footing.
11 October	The Pentagon orders I MEF to deploy its headquarters staff to Kuwait for service with Coalition Forces Land Component Command (CFLCC) under U.S. Army LtGen David D. McKiernan.
15 November	I MEF headquarters deploys to Kuwait; newly appointed I MEF commander LtGen James T. Conway deploys with his headquarters.
16 November	3d Marine Aircraft Wing (3d MAW) forward command post, under MajGen James F. Amos, arrives in Kuwait.
18 November	1st Marine Division forward command post arrives in Kuwait.
24 November	CFLCC exercise to test command and control links with I MEF and other commands, "Lucky Warrior 03-1," begins.
9 December	CentCom exercise "Internal Look," based on the current version of the plan for the invasion of Iraq, begins.

2003

January	Intense preparations to integrate 1st Armoured Division (UK) into I MEF occur; this division assumes responsibility for securing southeast Iraq.
2 January	Pentagon issues Deployment Order 177A, soon to be followed by 177B, which orders the wholesale deployment of I MEF forces to theater.

6 January	Rehearsal of Concept (ROC) drill occurs at 3d MAW in Miramar, California; many ROC drills at various levels follow in the coming weeks.
13 January	Gen Michael W. Hagee becomes the 33d Commandant of the Marine Corps.
15 January	Amphibious Task Force (ATF) East departs Morehead City, North Carolina, for Kuwait with 2d Marine Expeditionary Brigade (2d MEB).
17 January	Amphibious Task Force (ATF) West departs San Diego, California, for Kuwait carrying elements of I MEF.
8 February	With I MEF, participation, CFLCC exercise "Lucky Warrior 03-2," labeled "a dress rehearsal" for war, begins.
16 February	2d MEB begins to go ashore in Kuwait to reinforce I MEF; its aviation elements transfer to 3d MAW control and the ground elements are redesignated Task Force Tarawa.
24 February	Amphibious Task Force West begins offloading its West Coast Marine units in Kuwait; most other Marines follow by air.
9 March	First leaflets dropped on Baghdad urging noninterference with Coalition operations and soliciting support from Iraqi people.
17 March	President Bush issues an ultimatum to Saddam Hussein to leave Iraq within 48 hours.
18 March	Operation Southern Watch aircraft conduct air strikes against Iraqi early warning radars and command-and-control capabilities; Marine forces are ordered to staging areas.
Night of 19-20 March	U.S. Air Force aircraft and Navy vessels conduct unplanned attack against Saddam Hussein and other Iraqi leadership targets in what becomes popularly known as the "decapitation strike," which does not succeed but does initiate hostilities
20 March	Iraq retaliates by firing surface-to-surface missiles against Coalition troops in Kuwait; ground combat operations begin at night; I MEF is supporting attack to Army's V Corps; Regimental Combat Team (RCT 5) is leading Marine unit.
21 March	Marines capture the Rumaylah oil fields, a key CentCom objective; Marines and British forces secure the port of Umm Qasr before moving on the city of Basrah, the most important British objective.
23 March	Task Force Tarawa begins to secure the city of An Nasiriyah and its key bridges over the Euphrates River and the Saddam Canal; heavy fighting ensues; friendly fire incident occurs at bridge over canal; II Marine Expeditionary Force commander MGen Henry P. Osman deploys to northern Iraq to establish the Military Coordination and Liaison

	Command (MCLC) under operational control of CentCom in order to maintain political stability.
Night of 24–25 March	"Mother of all sandstorms" begins, slowing operations' tempo for approximately two days.
24–27 March	1st Marine Division continues to advance up Routes 1 and 7 towards Baghdad.
27 March	Operational pause begins to consolidate supply lines and address threats by irregular Iraqi formations on the ground; 3d MAW air offensive continues unimpeded, rendering many Iraqi units combat ineffective.
1 April	1st Marine Division resumes progress towards Baghdad; 1st Force Service Support Group performs herculean feats of resupply with cooperation of wing and Marine Logistics Command.
3 April	U.S. Army troops move on Saddam International Airport, key terrain outside Baghdad.
5 April	U.S. Army conducts first "Thunder Run," armored raid, into Baghdad.
6 April	Most of Basrah, Iraq's "second city," is in British hands.
7 April	Regimental Combat Team 7 (RCT 7) crosses the Diyala River and moves on outskirts of Baghdad from the east; U.S. Army conducts second "Thunder Run" into capital.
9 April	Marines of 3d Battalion, 4th Marines, part of RCT 7, assist Iraqi civilians in toppling a large statue of Saddam Hussein in Firdos Square in Marine area of operations, eastern Baghdad.
10 April	RCT 5 engaged in heavy fighting at Al Azimilyah Palace and Abu Hanifah mosque in Baghdad; looting begins as fighting tapers off; Marines begin post-combat operations.
11–12 April	After the collapse of Iraqi authority in northern cities of Mosul and Kirkuk, Kurdish forces fill the resulting power vacuum, followed by U.S. forces over succeeding days, including Marines from 26th MEU (SOC).
13-14 April	Task Force Tripoli, out of 1st Marine Division, takes control of Tikrit, Saddam Hussein's hometown.
20 April	The relief in place with U.S. Army in eastern Baghdad is complete; I MEF redeploys its forces to the southern third of Iraq; mission is now security, humanitarian assistance, and reconstruction; focus of effort is seven infantry battalions from 1st Marine Division in seven governates or districts.
22 April	24th MEU (SOC), which had supported Task Force Tarawa, begins

	redeploying to its ships; other Marine units soon follow suit as part of drawn-down to reduced manning levels that are maintained throughout the summer.
1 May	Under a banner reading "Mission Accomplished," President George W. Bush announces that major combat operations are over; 26th MEU (SOC) departs Mosul and returns to its ships in the Mediterranean.
12 May	Ambassador L. Paul Bremer takes over as civil administrator in Iraq, replacing Jay M. Garner; Bremer's Coalition Provisional Authority soon replaces Garner's Office of Reconstruction and Humanitarian Assistance.
22 July	Saddam Hussein's sons Uday and Qusay are killed in firefight with U.S. Army in Mosul.
19 August	A truck bomb explodes at the U.N. headquarters in Baghdad, killing 20 people, including the U.N. High Commissioner for Human Rights.
3 September	In Babylon, I MEF conducts a transfer of authority to a Polish-led international Coalition force; most remaining Marines return to Continental United States.
10 November	Marines of Special Purpose MAGTF celebrate the Marine Corps birthday in Continental United States after completing the work of repatriating all Marine Corps equipment from theater.

Appendix F

Presidential Unit Citation

THE SECRETARY OF THE NAVY
WASHINGTON, D.C. 20350-1000
3 November 2003

The President of the United States takes pleasure in presenting the PRESIDENTIAL UNIT CITATION to

I MARINE EXPEDITIONARY FORCE

for service as set forth in the following

CITATION:

For extraordinary heroism and outstanding performance in action against enemy forces in support of Operation IRAQI FREEDOM from 21 March to 24 April 2003. During this period, I Marine Expeditionary Force (MEF) (REIN) conducted the **longest sequence** of coordinated combined arms overland attacks in the history of the Marine Corps. From the border between Kuwait and Iraq, to the culmination of hostilities north of Baghdad, I MEF advanced nearly 800 kilometers under sustained and heavy combat. Utilizing the devastating combat power of organic aviation assets, coupled with the awesome power resident in the ground combat elements, and maintaining momentum through the herculean efforts of combat service support elements, I MEF destroyed nine Iraqi Divisions. This awesome display of combat power was accomplished while simultaneously freeing the Iraqi people from more than 30 years of oppression and reestablishing basic infrastructure in the country. During the 33 days of combat, to the transition to civil-military operations, I MEF sustained a tempo of operations never before seen on the modern battlefield, conducting four major river crossings, maintaining the initiative, and sustaining forces. The ferocity and duration of the campaign was made possible through the skills and determination of the Soldiers, Sailors, Airmen, Marines, and Coalition Partners comprising I MEF at all levels, all echelons, and in all occupational fields. By their outstanding courage, aggressive fighting spirit, and untiring devotion to duty, the officers and enlisted personnel of I Marine Expeditionary Force (REIN) reflected great credit upon themselves and upheld the highest traditions of the Marine Corps and the United States Naval Service.

For the President,

Gordon England

Secretary of the Navy

Index

Made in the USA
Middletown, DE
05 November 2026

36546287R10113